Sustainable Machining Using MQL Application of Cutting Fluids

The application of metal cutting fluids is an integral part of industrial machining operations. Minimum quantity lubrication (MQL) is the latest form of cutting fluid application method currently used by several manufacturing organizations. This book consolidates all the available knowledge in terms of the application of different processes as well as materials in a concise fashion in one reference resource.

Sustainable Machining Using MQL Application of Cutting Fluids offers a detailed discussion of the MQL mechanism in cutting fluid applications. It highlights the influence of MQL parameters on different workpiece materials and provides sound explanations along with photographs for all technical reasonings. The book presents the usage of both micro and nano cutting fluids in machining for sustainability while it also captures the knowledge in the field including the recent research outputs, as it illustrates a comprehensive coverage of MQL practical application.

This book should be on the bookshelf of industrial engineers, those working in production and manufacturing, process designers, tool material designers, cutting tool designers, and quality specialists. Researchers, senior undergraduate students, and graduate students will also find this book full of very helpful reference information.

Sustainable Machining Using MQL Application of Cutting Fluids

Nageswara Rao Posinasetti
Vamsi Krishna Pasam
Rukmini Srikant Revuru
Basil Kuriachen

CRC Press
Taylor & Francis Group
Boca Raton London New York

CRC Press is an imprint of the
Taylor & Francis Group, an **Informa** business

Designed cover image: www.shutterstock.com

First edition published 2024
by CRC Press
6000 Broken Sound Parkway NW, Suite 300, Boca Raton, FL 33487-2742

and by CRC Press
4 Park Square, Milton Park, Abingdon, Oxon, OX14 4RN

CRC Press is an imprint of Taylor & Francis Group, LLC

© 2024 Nageswara Rao Posinasetti, Vamsi Krishna Pasam, Rukmini Srikant Revuru, and Basil Kuriachen

ISBN: 978-1-032-35689-1 (hbk)
ISBN: 978-1-032-35820-8 (pbk)
ISBN: 978-1-003-32874-2 (ebk)

DOI: 10.1201/9781003328742

Typeset in Times
by MPS Limited, Dehradun

Contents

Preface

From the beginning of the Industrial Revolution, cutting fluids played a major role in improving the productivity of metal cutting operations. The use of cutting fluids is termed *metal working fluids* is to improve the quality of the machined components along with improving the life of the cutting tools. In addition to these main uses, they also serve a number of purposes. As a result, these are an essential part of any machining complex. It is reported that in German automotive companies, the average cost of cutting fluids is as high as 7 to 17% compared to the cutting tool cost that is 2 to 4%. Verified Market Research reports that the global market size of the metal working fluids is valued at US$9.91 billion in 2020 and is projected to reach US$12.63 billion by 2028, growing at a CAGR of 3.10% from 2021 to 2028.

Since the cutting fluids generate mist due to the heat produced during machining, the machine operators are exposed to these harmful chemical substances continuously over long periods. The accumulation of mist in the lungs of the operators is responsible for severe respiratory problems for long time machine operators. In addition, skin cancer is prevalent among the long-term machine tool operators. As a result, reducing the harmful metal working fluids is a necessity among the major metal working industries.

Further, the effectiveness of cutting fluids degrades over time and this eventually calls for disposal of the spent fluids. Because of their complex formulations, regular cutting fluids pose problems in their recycling and disposal. Increased environmental awareness and stringent regulations across the world, makes the disposal of cutting fluids an expensive proposition. Many companies are forced to spend on the cleanups due to poor waste disposal strategies. Because of their environmental impact, the use of extreme pressure additives, biocides, and other additives are often restricted.

Thus, alternative solutions for the reduction in the use of normal cutting fluids is a necessity to reduce the fluid costs as well as moving towards sustainability. The application of low quantities of cutting fluids is currently termed *minimum quantity lubrication* (MQL). The concept of MQL arises from the fact that a majority of the cutting fluid used in flooding is not really utilized for any purpose other than flushing chips and splashing. The idea of MQL is not new and a lot of interest was demonstrated in the 1960s and 1970s when it was called *mist cooling*. The main purpose here is to use as little cutting fluid as required that provides just enough cooling and lubrication for the required situation. General range of flow rate in MQL is of the order of 5 to 500 mL/hr. This is far less than the conventional quantities, which are of the order of 120,000 mL/hr. This low volume of fluid is generally supplied with the assistance of compressed air.

What has been observed is that with such a small amount of fluid, it is possible to achieve better machining performance using MQL by carefully controlling the process parameters. Thus, MQL would naturally lead to sustainability by reducing the amount of cutting fluid used in the application, thereby reducing the need for maintaining and disposing the used fluid as all of it is vaporized during the application.

Another dimension to the sustainability is the use of vegetable fluids as a base oil for cutting fluids, making them biodegradable and less harmful to the environment. A lot of interest in research is leading to various vegetable fluids being experimented for this purpose. It is noticed that there is a large number of publications on MQL and cutting fluids during the past decade. A leading automobile manufacturer equipped with an MQL system for its cylinder head machining has reduced lubricant consumption by 98%, water consumption by 90% and energy consumption by 54%, resulting in 46% lower CO_2 emissions. Ford Motor Company has started using MQL on more than 400 of their CNC machining centers worldwide to machine the engine and transmission components.

Thus, MQL is slowly becoming mainstream with a number of machine shops. The adoption of MQL will help a lot if all the knowledge for its adoption is made available in a single place with all the data for the process and parameter selection identified. The aim of this book is precisely that purpose. It organizes the available knowledge in terms of application to situations for different processes as well as materials. This book also consolidates the research in such a way that all the knowledge that is generated from this vast pool of research is organized so that the future researchers will be able to properly focus their research work in the knowledge gaps so that MQL will become more mainstream compared to today.

Nageswara Rao Posinasetti,
Vamsi Krishna Pasam,
Rukmini Srikant Revuru,
Basil Kuriachen

Author bios

Dr. P. Nageswara Rao is currently working as a professor in the Department of Applied Engineering and Technical Management, University of Northern Iowa, Cedar Falls, Iowa. He has 50 years of experience in teaching and research. His active areas of teaching and research are manufacturing engineering, design engineering, sustainability, Industry 4.0, smart manufacturing, and MEMS/nano technology education. He has wide interaction with the industry through the process of consultancy work and conducting continuing education programs on various aspects related to modern manufacturing. He has received a number of grants from various bodies to support his research activities. He was a recipient of the "Distinguished Scholar" award from the University of Northern Iowa for the years 2017–2018 and Fulbright Scholar in the year 2022. He has authored a number of textbooks on manufacturing technology, CAD/CAM, and metal casting, published by McGraw Hill India and American Foundry Society. He has also published over 260 research papers in international conferences and journals.

Dr. P. Vamsi Krishna is currently working as Professor in the Department of Mechanical Engineering, National Institute of Technology Warangal, India. He is having 21 years of experience in Teaching and Research. His research interests are application of solid lubricants in machining process, application of eco-friendly nano cutting fluids in machining, vibration assisted machining, modeling and simulation of manufacturing processes and composite materials. Five students were awarded Ph.D and five are pursuing Ph.D under his guidance. He guided 35 PG and 19 UG projects. He Completed five funded projects and one is ongoing. He published one text book, three text book chapters and edited three books. He published more than 130 research papers in reputed journals and conferences and one patent. He is editorial board member for 10 journals and delivered 35 guest lectures. He is recognized as one of the top 2% most influential scientists in 2022 Stanford university list. He is awarded with outstanding reviewer award by Tribology international, Elsevier in 2017 and best reviewer award from Transactions of Indian Institute of Metals in 2020.

Dr. Rukmini Srikant Revuru is currently working as an associate professor in the Department of Applied Engineering and Technical Management, University of Northern Iowa, USA. He currently has about 20 years of experience in teaching and research. His research interests include machining, cutting fluids, artificial intelligence, and robotics. He has completed six funded projects and was awarded a patent by the Indian patent office. He has published over 80 papers in various journals and conferences.

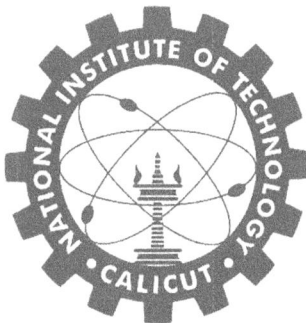

Dr. Basil Kuriachen is an assistant professor in the Department of Mechanical Engineering, National Institute of Technology Calicut. He has eight years of experience in teaching and research. His vivacity and dexterity towards abiding commitment to sublime work ethic has conferred him with the PhD from NIT Calicut (2015). His resolute research niches are in the field of metal additive manufacturing, micro/nano-machining processes, and precision and ultra-precision machining. He is the principal investigator (PI) for many currently active projects from DST-SERB, ARDB, DRDO, Govt. of India, UGC, DAE, TIH_IITG, SERB-IMPRINT, and ISRO. His hard work and dedication in academics and research enabled him to be the recipient of several prestigious national awards/fellowships, including the IEI Young Engineers Award 2019–20, SERB-International Travel Fellowship 2017, Early Career Research Award of DST-SERB 2016, and SERB's International Research Experience (SIRE) fellowship. He has, to his credit, 130 research publications in international refereed journals and conferences alongside two patents. He has authored a textbook, *Electric Discharge Hybrid-Machining Processes: Fundamentals and Application,* published by CRC Press.

1 Introduction

1.1 IMPORTANCE OF CUTTING FLUIDS FOR PRODUCTIVITY

In the early 20th century F. W. Taylor used water as a means of cooling the machining process that resulted in increased tool life. Since that time a large variety of cutting fluids have been used to remove heat as well as reduce the heat generation during the machining operations (Shaw, 2005). The use of cutting fluids greatly enhances the machining quality, at the same time reducing the cost of machining. In view of the variety of materials and machining situations, a large number of cutting fluids have been developed and used.

Cutting fluids will reduce the heat from the tool and the workpiece, and also lubricate the chip tool interface thereby reducing the friction. Some other mechanisms for the reduction in the shear strength of the work material used in the machining process was suggested by Russian authors that involves the presence of micro-cracks on the surface of the workpiece material. The overall impact of all this is that the energy consumed in the operation gets reduced while improving the life of the tool as well as the quality of the part surface produced. In addition, there are a number of other associated benefits that are difficult to quantify, but nevertheless are required in a machining operation. As a result, the number and quantity of cutting fluids used has increased over the past century to accommodate the range of materials developed.

In the beginning, cutting fluids are thought to improve the cutting performance, mostly because of the cooling properties. Hence, they were termed *coolants*. Also, higher temperatures increase the tool wear, and cooling of the cutting tool helps in maintaining the original properties of the cutting tool thereby prolonging its life. But it is a known fact that at lower temperatures, the shear flow stress of the workpiece may increase, thereby decreasing the tool life (Seah et al., 1995). The fact is that a number of investigations have actually demonstrated that cooling is in fact a major factor in improving the cutting performance.

However, the lubricating action will provide a better improvement in cutting fluid usage. The friction at the chip tool interface will be reduced thereby reducing the energy input to the metal cutting operation and consequently the heat generated (11 to 20%). In order for this to happen, the cutting fluid should reach the chip tool interface. As the chip-tool contact pressure approaches to a high value such as 70 MPa, it is extremely difficult for the cutting fluid to reach the chip-tool interface during a high-speed operation, such as turning tool. It has been suggested by Merchant (1950) that the cutting fluid will move through the asperities present in the chip-tool interface by the capillary action.

During the past century, organic and inorganic materials have been used to develop a variety of cutting fluids. However, in view of the environmental effects of

DOI: 10.1201/9781003328742-1

1

these cutting fluids, modern manufacturing industries are forced to pay greater attention to the sustainability. During the use of cutting fluids, mist and vapor are generated because of the heat generated by the machining processes, which is harmful for the operator. Therefore, stringent regulations have been developed to control them. It has been noticed that direct exposure of cutting fluids to the machining operators is responsible for a number of skin cancer cases (Calvert et al., 1998). Also, the disposal of spent cutting fluids should be recycled or disposed of in a manner that is not harmful to the environment. To this effect, stringent environmental legislations are available. This means that extra cost to the manufacturing organizations in the recycling and disposal procedures of the cutting fluids depending upon the type of cutting fluid.

Klocke and Eisenblatter (1997) have reported that an estimated 350,000 tonnes of cutting fluids were processed and subsequently disposed of in Germany alone in 1994. The actual cost of purchasing the cutting fluid and its disposal is about 1 billion German marks (Dry Machining's Double Benefit, 1994). According to the German automotive industry, the tool costs are quoted in the range of 2–4% while cutting fluids cost is 7–17% of the total manufacturing costs for the machining of the automobile components. This is in view of the associated costs of monitoring, maintenance, health precautions, and absenteeism in automobile industry. Hence, there is a need to increase the emphasis on the research that can lead to the reduction in the costs associated with the cutting fluids by reducing the cutting fluid volumes used as well as recycling and disposal methods.

In Japan, it is reported that in 1984, the manufacturing industry has spent about 29 billion Japanese yen a year towards the purchase of cutting fluids (Feng & Hattori, 2000). It further specified that the majority is water-immiscible fluids of 100,000 kiloliters, water-soluble coolant without chlorine of 50,000 kiloliters, and water-soluble coolant with chlorine of 10,000 kiloliters. The disposal cost is 35–50 yen per liter for the water-immiscible coolant, 300 yen per liter for water-soluble coolant without chlorine, and 2,250 yen per liter for water-soluble coolant with chlorine. Based on the above figures, the estimated coolant disposal cost alone in Japan is about 42 billion yen. The total coolant purchasing and disposal cost is about 71 billion yen a year (Feng & Hattori, 2000).

1.2   HISTORICAL DEVELOPMENTS IN CUTTING FLUIDS AND METAL CUTTING

Though metals have been used for long periods before the Industrial Revolution, there is the use of lubricants in general for the metal working process. Albert et al. (2014) suggested that it may not be unreasonable to presume that the lubrication then was used with the oils that are readily available. The fluids that were readily available at that time include the vegetable oils such as olive, palm, castor, and other seed oils as well as animal oils and fats (primarily whale, tallow, and lard) (Dowson, 1979). The early reference to the use of metal working fluids is in an 1868 book by Northcott *A Treatise on Lathes and Turning,* though it was already widespread in England and the United States. Browne and Sharpe presented a device for carrying off water or other metal working fluids for the grinding machines in their 1875 catalog. The use of sulphur as an additive in oils for easing the machining operations was noted around 1882. It was found during that time that sulphurized oils were good for difficult-to-machine materials by cooling and reducing the adhesion of chips to the tools.

Taylor (1906) conducted a large number of machining experiments to improve the machining productivity. His technique was to direct flooding of the cutting edge by water with a rust inhibitor (sodium bicarbonate). By 1900, cutting fluids were being circulated continuously in the machine tools to remove the heat from the cutting zone with water or oil-based cutting fluids.

With the widespread availability of petroleum, the by-products during the extraction of useful products from petroleum have been found to be useful as lubricants when combined with fatty oils to act as metal working fluids. With many developments in hard alloy steels during the early 20th century, many compounded cutting fluid combinations by combining sulfur and chlorine to cater to the different metallurgy of the new materials that were made available.

It was realized that the environmental degradation due to human activity was progressing at an alarming rate, and sustainability of the biosphere was an important requirement at the current stage. It was therefore imperative that humans should manage their activities judiciously so that sustainability of the planet became a core concern for all. To this extent, metal cutting activity has also progressed towards sustainability by adopting renewable cutting fluids.

1.3 PROBLEMS WITH CUTTING FLUIDS

As the cutting fluid is continuously used over time, its composition in the shop is likely to change unless proper care is taken. This change will occur from two different sources:

- Regular changes in the cutting fluid over time due to the use of constituents; and
- Contamination of the cutting fluid from a variety of external sources.

It is generally understood that there will be an increase in concentration of alkanol amines in the system over time, relative to other components. Also, the constant addition of water to the fluid circulating system increases the amount of metal salts, particularly when hard water is used. These salts tend to destabilize semi-synthetic and soluble oils. With use, corrosion inhibitors and biocides decrease over time. It is therefore necessary to regularly add these components to keep the fluid performance at the desired level by following a regular program of fluid testing and analysis.

With regular use, industrial lubricants contaminate the circulating cutting fluid systems by direct contact. This is called "tramp oil", and may be contributed from leaking hydraulic fluid or other lubricants with which the cutting fluid is likely to come in contact during its flow through the machine tool structure. This tramp oil may be miscible with the cutting fluid, but is likely to alter the characteristics of the cutting fluid. Poorly maintained machine tools may contribute more tramp oil than the well-maintained machine tools.

Practically all water-soluble oils become contaminated with bacteria (Foxall-VanAken et al., 1986; Rossmore, 1993). The presence of bacteria degrades the various components of the fluid. Bacterial contamination of fluids provides a high risk of opportunistic pathogens (Rossmore et al., 1987). Endotoxins have been identified as a potential risk from exposure to metal removal fluids (Gordon, 1992). Fungal contamination appears to occur less frequently than bacterial contamination.

1.3.1 ROUTES OF EXPOSURE

During the use of metal working fluids, the operators are exposed to direct contact with the fluid and also inhale the mist. Generally, automating the machining work stations has reduced the requirement of the machining operator to be present in front of the machine tool, but it is still a problem for older machining equipment. In particular, the manual loading and operating machines do require that the operator be present and is then exposed to the metal working fluid for a greater part of time. The bare skin contacts with the metal working fluid causes dermatitis and this happens to be the most commonly reported medical situation (Bennett, 1983; Foulds & Koh, 1990). From 0.3 to 1% of machinists have been estimated to have either contact dermatitis or allergic contact dermatitis (Kennedy et al., 1989). For healthy persons, exposure to fluids only may not be the cause of dermatitis. There is a likelihood that they may have some other problem in the shop, such as any of the following that would cause skin problems:

- poor personal hygiene
- seasonal conditions
- contamination of the fluid by dissolved metals, abrasive particulates, or alkaline materials
- rich concentrations
- dirty shop rags
- filter malfunction
- use of solvents
- abrasive soaps
- off-job activities

In most of these cases, the fluid manufacturer, user, and medical personnel will try to manage such dermatitis situations by following the specified procedures to eliminate the route problems. Inhalation of mist and vapor is the second most common route of exposure, which can be controlled by following specific procedures and practices of the equipment used for the purpose.

Here are a few examples of the types of exposures noticed in different industrial plants. The total fluid mist exposure to the machine operators is 0.16 and 2.03 mg/m^3 in an automotive transmission facility, as found by Kennedy et al. (1989). They also noted that the exposure levels remain the same among all different types of metal working fluids. A similar result was obtained by Chan et al. (1990), who found between 0.71 and 2.99 mg/m^3 of total particulate concentrations in another automotive transmission facility. There are minor variations in different types of metal working fluids. In a statistical analysis conducted by Woskie et al. (1994a, 1994b), they found the average total particulate exposure to be 0.7 mg/m^3. They considered the variability in the particle size from small to large.

Kenyon et al. (1993) found that the metal working fluids used in the auto parts manufacturing industry contain ethanolamine in different percentages. For example, the metal working fluids contain from 1% to 11% of monoethanolamine (MEA), 4% to 5% diethanolamine (DEA), and 0.3% to 40% triethanolamine (TEA). They found that machining operators were exposed to higher amounts of TEA, while the assembly operators were less exposed since they did not use the machining fluids.

A survey of eight manufacturing plants by Ball (1995) found that the median exposure level was approximately 1 mg/m^3. Kennedy et al. (1989) are of the opinion that the allowable exposure levels to metal removal fluid aerosols were too high. They reached this conclusion based on the finding that a cross-shift decrease in forced expiratory volume with increasing exposure levels above approximately 0.20 mg/m^3.

1.4 SUSTAINABILITY OF CUTTING FLUIDS

The life cycle of the cutting fluids is shown in Figure 1.1. Most of the industrial metal working fluids are based on petroleum fluids. After properly preparing the cutting fluids, they are utilized for a period of time with appropriate topping methods employed to maintain the requisite properties. During the regular use for machining operations, as explained earlier, the operators were exposed to the fluid in the form of liquid or as mist, thus contributing to the various health issues. After the useful life of the cutting fluid is over, it then needs to be properly disposed of. If the spent cutting fluid is not properly disposed, it is likely to contaminate the soil

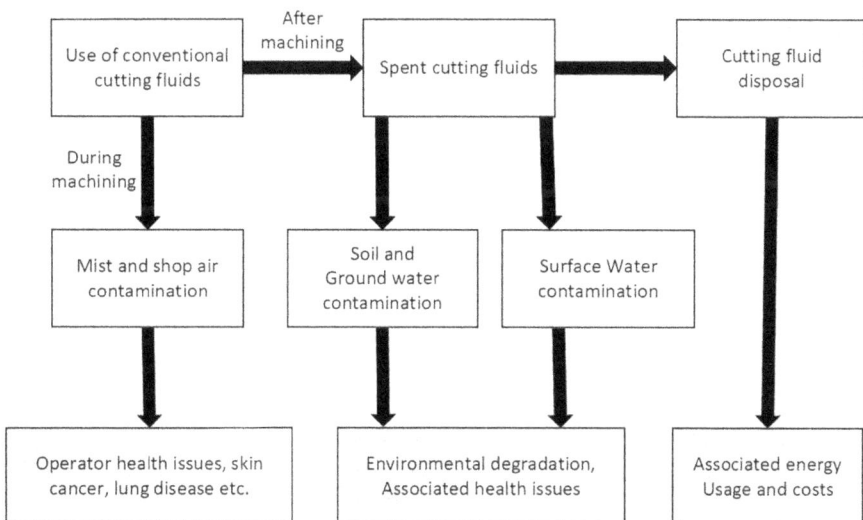

FIGURE 1.1   Life cycle of cutting fluids through a machine shop.

and water sources, which will lead to long-term problems for the local humanity. On the other hand, following the guidelines for proper recycling requires extra energy and associated costs. It is therefore important that the need is to get biodegradable cutting fluids that can reduce the disposal costs for the metal working fluids.

Sustainable or renewable cutting fluids are based on vegetable oils, which are biodegradable as well as renewable (Belluco & De Chiffre, 2002, 2004; Diniz & De Oliveira, 2004; Khan & Dhar, 2006; Rao & Zhang, 2009; Rusinko, 2007; Singh & Gupta, 2006; Siniawski & Bowman, 2009; Smith & Frazier, 2010; Zhang et al., 2012). The petroleum-based cutting fluids have a relatively low flash point (about 215°C), which generate a large amount of mist due to the high temperature at the workpiece-cutter interface, which is harmful to the machine operators. One of the ways this can be minimized is to utilize genetically modified soybean-based cutting fluids that have a high molecular weight and high flash point of around 315°C. A higher flash point ensures that the chance of mist generation in machining processes is reduced. Because of the very high film strength, these GM soy-based metal working fluids ensure that the cutting tool workpiece and chip interfaces have reduced friction reducing heat and tool wear (Zhang et al., 2012).

The widespread use of renewable and biodegradable cutting fluids in industries is not prevalent, though they have been in the market for some time. However, a few studies have been reported on the use of bio-based metal working fluids in the literature (Belluco & De Chiffre, 2002, 2004; Diniz & De Oliveira, 2004; Khan & Dhar, 2006; Rao & Zhang, 2009; Rusinko, 2007; Singh & Gupta, 2006; Siniawski & Bowman, 2009; Smith & Frazier, 2010; Zhang et al., 2012). One early research is reported by Belluco & De Chiffre (2002, 2004), focusing on the use of blended cutting fluids. They blended rapeseed oil and ester oil and then added additives of sulfur and phosphorous and conducted research on drilling of AISI 316L austenitic stainless steel. Their results indicate that tool life has increased, chip breaking improved, and tool wear and cutting forces were lowered with the use of their formulated cutting fluids compared to mineral oil in drilling. Rao & Zhang (2009) and Zhang et al. (2012) indicated that the use of genetically modified soybean-based metal working fluids have improved the cutting performance compared to petroleum-based fluids. They have noted that the main advantage of these renewable cutting fluids is that no harmful additives are added to improve the performance of these fluids. That will help in the procedures used for their disposal, ultimately resulting in lower disposal cost. Later chapters will be discussing the various options that were developed that lead to the use of renewable cutting fluids, thereby reducing the carbon footprint for the cutting fluids.

REFERENCES

Albert, S., Choudhury, I. A., Sadiq, I. O., & Adedipe, O. (2014). Vegetable-oil based metalworking fluids research developments for machining processes: Survey, applications and challenges. *Manufacturing Rev*, 1(22), 1–11. (DOI: 10.1051/mfreview/2014021)

Ball, A. M. (1995). A survey of metalworking fluid mist in manufacturing plants. Paper presented at the 50th Annual Meeting of the Society of Tribologists and Lubricating Engineers, Chicago, IL, May.

Belluco, W., & De Chiffre, L. (2002). Surface integrity and part accuracy in reaming and tapping stainless steel with new vegetable based cutting oils. *Tibology International*, 35(12), 865–870.

Belluco, W., & De Chiffre, L. (2004). Performance evaluation of vegetable-based oils in drilling austenitic stainless steel. *Journal of Materials Processing Technology*, 148(2), 171–176.

Bennett, E. O. (1983). *Dermatitis in the metalworking industry*. Park Ridge, IL: Society of Tribologists and Lubrication Engineers, [Pamphlet SP-11].

Calvert, G. M., Ward, E., Schnorr, T. M., & Fine, L. J. (1998). Cancer risks among workers exposed to metalworking fluids: A systematic review. *American Journal of Industrial Medicine*, 33, 282–292.

Chan, T. L., D'Arcy, J. B., & Siak, J. (1990). Size characteristics of machining fluid aerosols in an industrial metalworking environment. *Applied Occupational and Environmental Hygiene*, 5(3), 162–170.

Diniz, A. E., & De Oliveira, A. J. (2004). Optimizing the use of dry cutting in rough turning steel operations. *International Journal of Machine Tools and Manufacture*, 44(10), 1061–1067.

Dowson, D. (1979). *History of tribology* (pp. 177–178). New York: Longmans Green, 253.

"Dry Machining's Double Benefit," *Machinery and Production Engineering*, June 1994, pp 14–20.

Feng, S. C., & Hattori, M. (Oct. 2000). Cost and process information modeling for dry machining. *Proceedings of the International Workshop on Environment and Manufacturing*.

Foulds, I. S., & Koh, D. (1990). Dermatitis from metalworking fluids. *Clinical and Experimental Dermatology*, 15, 157–162.

Foxall-VanAken, S., Brown Jr, J. A., Young, W., Salmeen, I., McClure, T., Napier Jr, S., and Olsen, R. H. (1986). Common components of industrial metal-working fluids as sources of carbon for bacterial growth. *Applied and Environmental Microbiology*, 51(6), 1165–1169.

Gordon, T. (1992). Acute respiratory effects of endotoxin-contaminated machining fluid aerosols in Guinea pigs. *Fundamental and Applied Toxicology*, 19, 117–123.

Kennedy, S. M., Greaves, I. A., Kriebel, D., Eisen, E. A., Smith, T. J., & Woskie, S. R. (1989). Acute pulmonary responses among automobile workers exposed to aerosols of machining fluids. *American Journal of Industrial Medicine*, 15, 627–641.

Kenyon, E. M., Hammond, S. K., Shatkin, J., Woskie, S. R., Hallock, M. F., & Smith, T. J. (1993). Ethanolamine exposures of workers using machining fluids in the automotive parts manufacturing industry. *Applied Occupational and Environmental Hygiene*, 8(7), 655–661.

Khan, M. M. A., & Dhar, N. R. (2006). Performance evaluation of minimum quantity lubrication by vegetable oil in terms of cutting force, cutting zone temperature, tool wear, job dimension and surface finish in turning AISI-1060 steel. *Journal of Zhejiang University: Science*, 7(11), 1790–1799.

Klocke, F., & Eisenblatter, G. (1997). Dry cutting. *Annals of CIRP*, 46(2), 519–526.

Merchant, E. M. (1950). The action of cutting fluids in machinery. *Iron and Steel Engineering*, 27, 101–108.

Rao, P. N., & Zhang, J. Z. (2009). Green manufacturing—environmental effects of soy based cutting fluid. Presented at the ATMAE Convention, November, 2009.

Rossmore, H. W. (1993). Biostatic fluids, friendly bacteria, and other myths in metalworking microbiology. *Lubrication Engineering* 49(4), 253–260.

Rossmore, H. W., Rossmore, L. A., & Young, C. E. (1987). Microbial ecology of an automotive plant. In G. C. Llewellyn, & C. E. O'Rear (Eds.), *Biodeterioration Research 1* (pp. 255–268). New York: Plenum Publ.

Rusinko, C. A. (2007). Green manufacturing: An evaluation of environmentally sustainable manufacturing practices and their impact on competitive outcome. *IEEE Transactions on Engineering Management*, 54(3), 445–454.

Seah, K. H. W., Li, X., & Lee, K. S. (1995). The effect of applying coolant on tool wear in metal machining. *Journal of Materials Processing Technology*, 48, 495–501.

Shaw, M. C. (2005). *Metal cutting principles*, 2nd ed. London: Oxford University Press.

Singh, A. K., & Gupta, A. K. (2006). Metalworking fluids from vegetable oils. *Journal of Synthetic Lubrication*, 23(4), 167–176.

Siniawski, M., & Bowman, C. (2009). Metal working fluids: Finding green in the manufacturing process. *Industrial Lubrication and Tribology*, 61(2), 60–66.

Smith, D., & Frazier, B. (2010). The benefits of bio-based lubricants. *Gear Solutions*. Retrieved from http://www.gearsolutions.com/article/detail/5990/the-benefits-of-bio-based-lubricants.

Taylor, F. W. (1906). *On the art of cutting metals* (Vol. 23). New York: American Society of Mechanical Engineers.

Woskie, S. R., Smith, T. J., Hallock, M. F., Hammond, S. K., Rosenthal, F., Eisen, E. A., Kriebel, D., & Greaves, I. A. (1994a) Size-selective pulmonary dose indices for metalworking fluid aerosols in machining and grinding operations in the automobile manufacturing industry. *American Industrial Hygiene Association Journal*, 55(1), 20–29.

Woskie, S. R., Smith, T. J., Hammond, S. K., & Hallock, M. F. (1994b). Factors affecting worker exposures to metal-working fluids during automotive component manufacturing. *Applied Occupational and Environmental Hygiene*, 9(9), 612–620.

Zhang, J. Z., Rao, P. N., & Eckman, M. (2012). Experimental evaluation of a bio-based cutting fluid using multiple machining characteristics. *International Journal of Modern Engineering*, 12(2), 35–44. Spring/Summer 2012.

2 Metal cutting processes using cutting fluids

2.1 METAL CUTTING

The metal cutting process is one of the most complex processes. A typical cutting tool in a simplified form is shown in Figure 2.1 removing material in the form of a chip during a cylindrical operation. The cutting process parameters, cutting speed, feed, and depth of cut, have been shown that will affect the quality of the machining operation. For the purpose of understanding the mechanics of the metal cutting, the cutting zone is simplified and magnified, as shown in Figure 2.2, for our discussion.

Figure 2.2 shows the basic material removal operation schematically in a magnified form. The metal in front of the tool rake face gets immediately compressed first elastically and then plastically. This zone is traditionally called as the shear zone in view of the fact that the material in the final form will be removed by shear from the parent metal. The actual separation of the metal starts from the cutting tool tip as yielding or fracture, depending upon the cutting conditions. Then the deformed metal (called chip) flows over the tool (rake) face. If the friction between the tool rake face and the underside of the chip (deformed material) is considerable, then the chip becomes further deformed, which is termed *secondary deformation*. The chip, after sliding over the tool rake face, will be lifted away from the tool, and the resultant curvature of the chip is termed *chip curl*.

In metal cutting, plastic deformation can be caused when the material crosses the yield point, at which point the strained layers of material in front of the tool tip get displaced over other layers along the slip-planes. The direction of the slip-planes coincides with the direction of maximum shear stress. As the plastic deformation takes place in the shear zone, the friction mechanism at the contacting surfaces is different. Here, because of the acting load, the real area of contact approaches that of apparent area of contact (Armarego & Brown, 1969; Boothroyd & Knight, 1989; Shaw, 2005; Trent & Wright, 2000; Taylor, 1907). Microscopic examination of the cutting tool and chip (material removed from the workpiece) is seen; the actual contact of the two sliding surfaces has contact through the high spots (asperities). In view of this, the actual contact surface between the two is small because of these asperities.

Very high temperatures are developed at the contacting asperities because of the high normal and tangential load. This results in metallic bonding of the contacting high spots. Thus, sliding of the chip relative to the cutting tool must be accompanied by shearing of the welded asperities. Thus, the friction along the rake face of a cutting tool can be considered as partially sticking and partially sliding, as shown in Figure 2.3. Work material's yield stress is the limit for the shear stress in the sticking zone. In the sliding zone, the normal Coulomb's laws of friction would

DOI: 10.1201/9781003328742-2

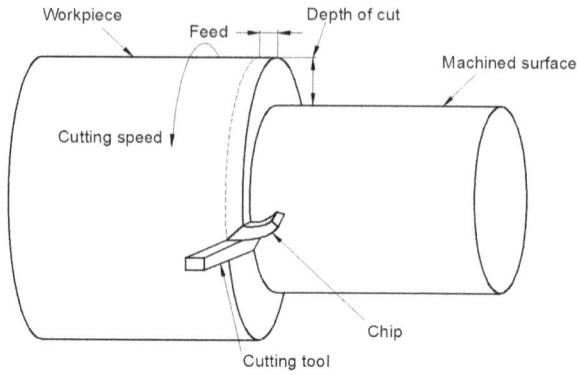

FIGURE 2.1   The cutting tool in action removing the work material in the form of chip during cylindrical turning operation.

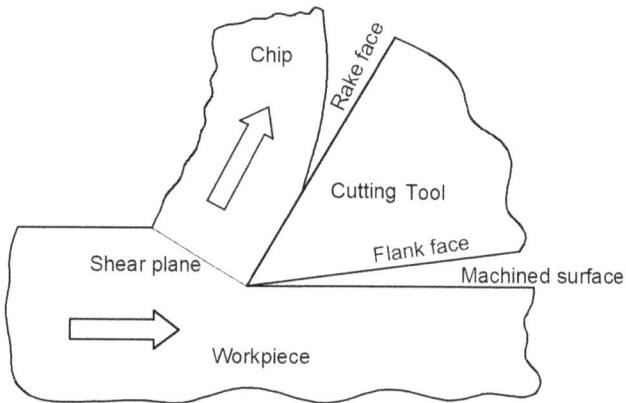

FIGURE 2.2   The cutting tool in action removing the work material in the form of chip in simplified form.

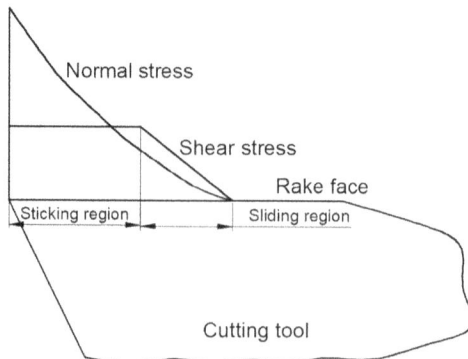

FIGURE 2.3   Friction conditions along the rake face of the cutting tool.

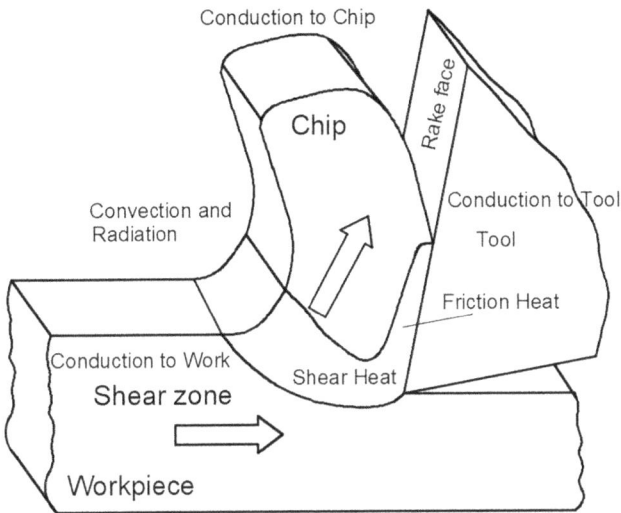

FIGURE 2.4   The main regions of heat generation in metal cutting along with the type of heat transfer mechanisms.

hold good. The combined effect of the sticking and sliding is that there will be a substantial amount of friction load between the chip and the cutting tool rake surface.

During metal cutting, less than 1% of the work done is stored as elastic energy. The remaining 99% of the energy is going to heat the workpiece, chip, and the cutting tool. Severe compression in the area ahead of the cutting edge of the cutting tool results in very high temperatures in the material in front of the cutting tool rake face. This results in plastic flow. It is understood that the internal friction of atoms sliding over one another because of severe loading may be causing the rise in the heat in the shear zone. The removed metal from the workpiece, which is termed a *chip,* slides on the rake face experiences the friction. Similarly, the sliding of the flank face with that of the finished surface also experiences the friction, providing a small component of heat (Figure 2.4).

Hence, the typical zones in metal cutting where the heat is generated (Figure 2.3) include the following:

A. This is the highest amount of heat generated of the order of 65 to 75% of the total heat generated. This part of the heat is generated because of internal friction and sliding of atoms in the shear zone.
B. Next, the major part is due to the friction at the chip tool interface. This causes a heat of the order of 15 to 25%, and
C. The smallest amount of heat of the order of 5 to 10% due to the friction at the tool work interface. (Figure 2.5)

As explained earlier, 99% of the shear energy would be converted into heat that would move into the cutting tool and the workpiece, thereby raising their temperatures. The

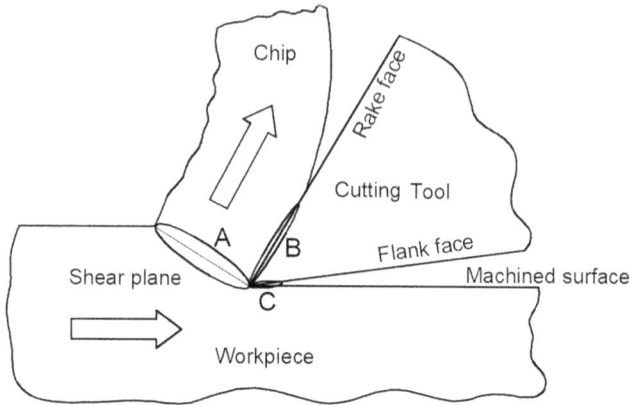

FIGURE 2.5   The main regions of heat generation in metal cutting.

component of heat that gets into the tool tip raises the temperature in that small area and remains in the cutting zone. The rest of the heat will be conducted into the chips and will be moved away from the cutting zone. A small amount of the heat will go back to the workpiece, thereby raising the temperature of the finished workpiece. The heat carried away by the chip will be partly lost to the surrounding atmosphere through convection and radiation, while the residual heat will be with the chips in the sump area.

The highest temperature will be seen at the tip of the cutting tool. The heat then would conduct through the tool along the rake face as well as the flank face, slowly in view of the lower thermal conductivity. As most of the tool wear mechanisms are thermally activated, the wear along the rake and flank faces of the tool will increase. To get a generic idea, Cook (1966) used dimensional analysis to develop an empirical formula to estimate the tool tip temperature as follows:

$$T = 0.4\, T_a \left(\frac{V\,d}{K}\right)^{\frac{1}{3}}$$

where K = thermal diffusivity of the work material, mm^2/s

c is the Specific heat;
ρ is the Density

Adiabatic temperature, $T_a = \frac{u_s}{\rho\,c}$
   Specific cutting energy, u_s given by

$$u_s = \frac{F_H\,V}{MRR} = \frac{\tau\,\cos(\beta - \alpha)}{\sin(\varphi)\cos(\varphi + \beta - \alpha)}$$

$$F_H = 2\,\tau\,b\,t\,\cot\varphi$$

where τ is the mean shear stress in the shear plane,

V is the cutting speed,
MRR is the material removal rate,
t is the chip thickness,
b is the width of cut,
α is the rake angle,
β is the friction angle, and
φ is the shear angle; $\varphi = \frac{\pi}{4} - \frac{1}{2}(\beta - \alpha)$

This equation is applicable to estimate the temperature for rake angles between −5 and +15°. Though a number of other analytical methods are developed and are available. However, the details of those analyses are beyond the scope of this book.

The extent of temperatures generated in a cutting tool are considerably affected by the cutting process parameters via cutting speed, feed, and depth of cut. The variable that has the highest effect in controlling the total energy input to the metal cutting operation is the cutting speed. The other variables are the feed and depth of cut in that order, in terms of their influence on the energy input in machining operation.

As the cutting tool is used over time, it will experience loss of material in the form of wear. The reason that the cutting tools experience this wear is because of the extremely severe conditions experienced at the chip tool interface, such as the following:

- virgin metal
- very high temperature
- very high stress
- metal to metal contact with workpiece and chip
- very high temperature gradients
- very high stress gradients

As a result, the tool chip and tool work interfaces exhibit gradual amount of loss of materials in two types at the rake face and the flank face. The type of wear found in cutting tools is shown in Figure 2.6.

The two major types of wear found in tools are as follows:

Crater wear: The crater wear found on the rake face is approximately circular. The crater wear starts at a distance from the tool tip. The effects of the presence of crater wear is that it increases the contact length of the chip, thereby promoting the cutting forces, modifies the tool geometry (increasing the rake angle) and softens the tool tip.

Flank wear: Flank wear, also called wear land, is on the clearance face of the cutting tool. It is characterized by the length of wear land, w. It modifies the tool geometry and changes the cutting parameters (depth of cut). It often affects the dimension produced and therefore important from the tool life point of view.

A detailed view of the crater wear is presented in Figure 2.7.

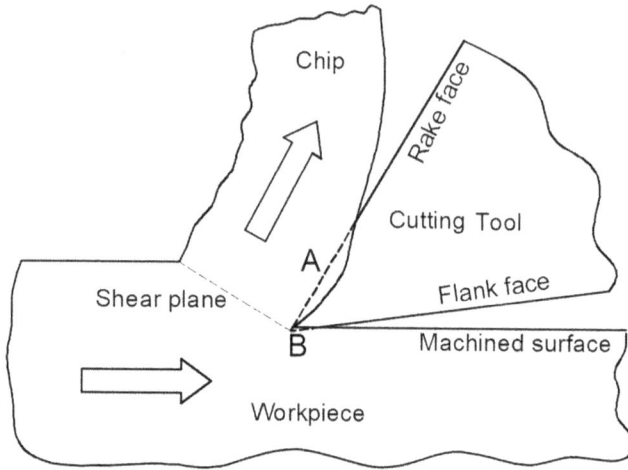

FIGURE 2.6 Wear regions in a metal cutting tool, A = Crater Wear, B = Flank Wear.

As cutting time progresses, tool wear increase that increases the cutting forces and vibrations. The tool tip softens and flows plastically and gets a blunt edge, which will result in further progressing of plastic deformation from tool tip to the interior. After that, the tip of the tool gets separated.

Relatively, crater wear is not important from the tool life point of view. The machining performance is not affected by the presence of small size crater wear. Small crater wear effect is to increase the rake angle, thereby improving the machining performance by lowering the cutting forces. But, when the crater wear progresses with the deepening crater, this would result in an increase in the friction and chip contact length. This likely deteriorates the machining performance. In rare cases, having a very large crater depth (KT) will weaken the tool tip, leading to a catastrophic failure. The cutting tool is normally replaced before this because it reached the flank wear limit.

Flank wear is more important for industrial applications because it directly affects the dimensions produced. The limit crater wear value is specified based on the dimensional tolerance required. The reason is that as both the crater wear and flank wear progress simultaneously, the tool tip becomes weak. Thus, in almost all instances, if the crater wear is allowed to go deep, then the cutting tool resharpening requires more time, as more material needs to be removed. This will reduce the number of regrinds possible for the tool tip.

ISO has suggested the tool life values, as follows (Figure 2.7):

$VB = 0.3$ mm if the flank is regularly worn in zone B, or

$VB_{max} = 0.6$ mm if the flank is irregularly worn, scratched, chipped, or badly grooved in zone B.

Tool wear is a complex phenomenon; a number of wear mechanisms were proposed to account for the complex behavior. They are:

- Adhesion
- Abrasion

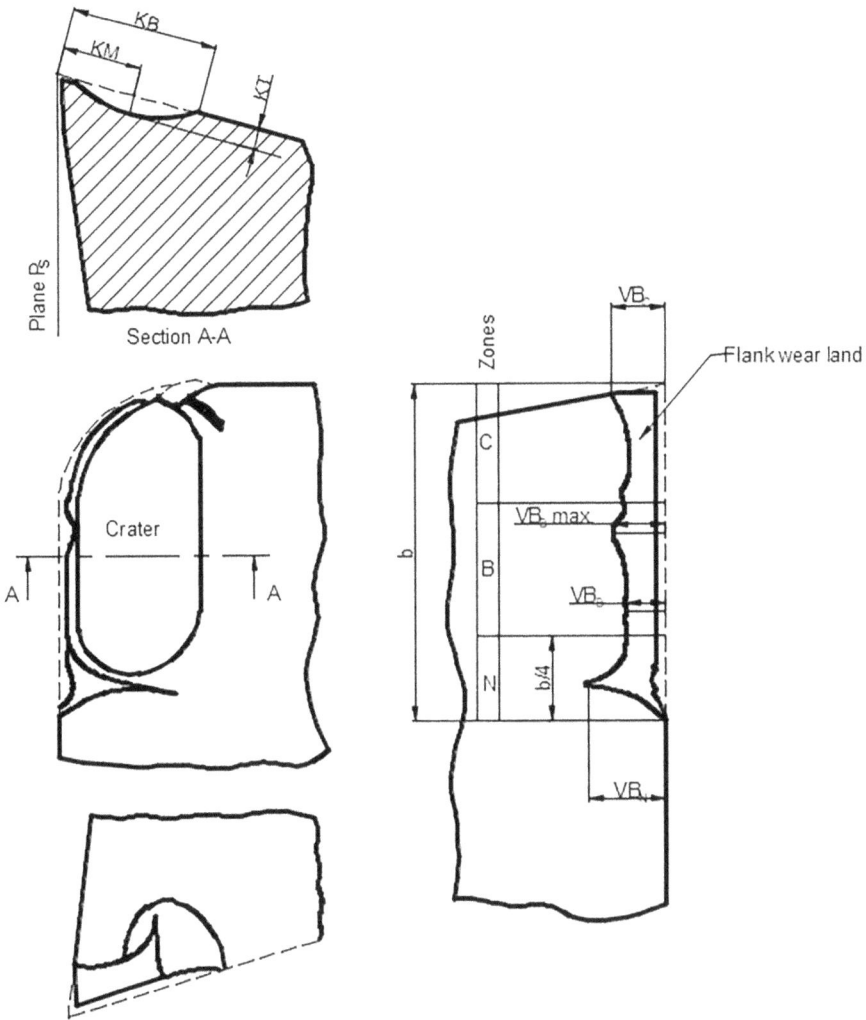

FIGURE 2.7 A detailed view of the tool wear regions and the parameters that are used for measurement.

- Diffusion
- Fatigue

Figure 2.8 illustrates the rate of change in crater wear patterns with cutting time. Figure 2.9 shows the rate of change in flank wear with cutting time. At low cutting speeds, the flank wear is more while at higher cutting speeds; both the wear rates are similar in quantitative terms. At very high cutting speeds, crater wear is the predominant wear that should be considered. This behavior may be because of the higher temperatures as all the wear mechanisms suggested earlier are thermally activated, causing the wear growth. Another aspect to be noted is that in the case of

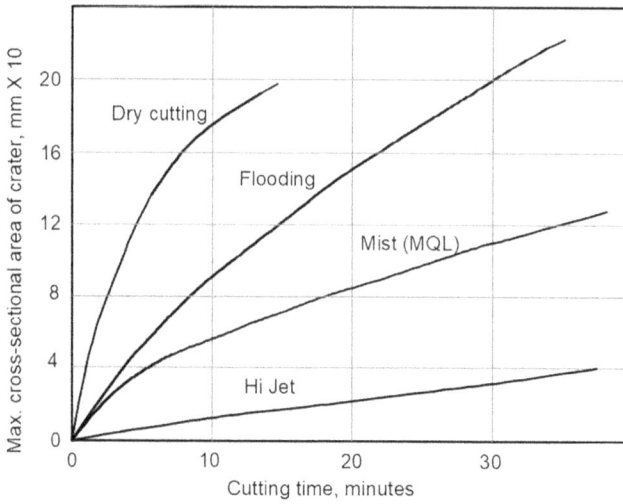

FIGURE 2.8   Crater wear progression with time (Rao, 1973; Rao & Arora, 1977).

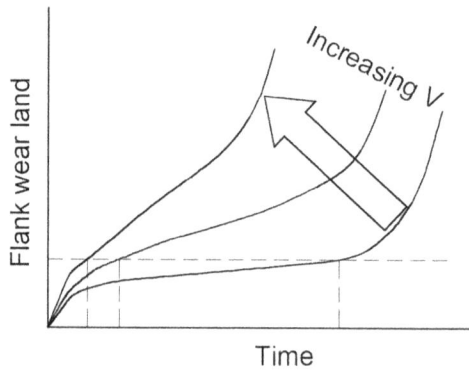

FIGURE 2.9   Flank wear progression with time as well as cutting speed.

flank wear, there is an initial rapid wear, while the crater wear curves are generally linear (Figure 2.8).

In Figure 2.9, the growth of flank wear can be seen as three distinct wear rates. The first stage is when the sharp cutting tool enters the workpiece to remove the material; all the heat generated would quickly raise the tool temperature to a very high value. This happens because the cutting tool material present at tool tip is very small that would not be able to conduct all the heat coming from the shear zone. Thus, the flank wear growth initially is very high, as can be seen in Figure 2.9 in the slope of the first region. However, when a small amount of flank wear is present, the area of the cutting tool area in contact with the workpiece becomes a little more and is able to conduct the heat at a higher rate, facilitating slow growth rate of the flank wear. This distributes heat uniformly to more volume of the tool thus the temperature gradually increases uniformly in the cutting tool. This is the second region in Figure 2.9. In the third region, the temperature rises catastrophically because of the increased friction and consequently the cutting forces. This finally brings the cutting tool to its end point of life. Also, the cutting speed increases the wear rate slopes increase, as shown in Figure 2.9.

2.2   CUTTING FLUID ACTION DURING MACHINING

There are many functions that are performed by the metal working fluids during a machining operation. They are:

- to cool the cutting tool and the workpiece to prolong the tool life
- to reduce the friction at the chip tool and workpiece tool interfaces
- to protect the workpiece against rusting
- to improve the surface finish of the workpiece
- to prevent the formation of built up edge (BUE) on the tool tip
- to wash away the chips from the cutting zone making the machining zone clearly visible for the operator.

Although these are all the functions performed by the cutting tool, the prime function of a metal working fluid in a metal cutting operation is to control the total heat that progresses into the workpiece and the cutting tool (Armarego & Brown, 1969; Boothroyd & Knight, 1989; Shaw, 2005; Trent & Wright, 2000; Taylor, 1907). This can be done by dissipating the actual heat generated in the shear zone while also reducing the heat generated due to the friction at the chip tool and workpiece tool interfaces. The mechanisms suggested by which a metal working fluid performs these functions may be listed as follows:

- Cooling action
- Lubricating action
- Rehbinder effect

In order for the cutting fluid to function properly, it should cool the cutting tool more effectively, which means that the fluid should enter machining zone properly. With the cutting force at the tool tip being extremely high, it is very difficult for the cutting fluid droplets to go beyond the tool tip into the chip tool interface in order to provide a reduction in the friction, thereby reducing the energy requirements for metal cutting. However, when the chip tool interface is seen under the microscope, the interface has the hills and valleys on both sides, as shown in Figure 2.10. It is suggested that the presence of these asperities provide interconnecting micro passages for the cutting fluid to penetrate by capillary action, thereby lubricating the chip tool interface. When the cutting fluid enters in any of the micro pores, the surface tension causes the cutting fluid to flow in the capillary against the motion of the workpiece surface and reach the tool tip. Here, because of the high temperature of the tool tip, the cutting fluid vaporizes, which facilitates the fluid penetration.

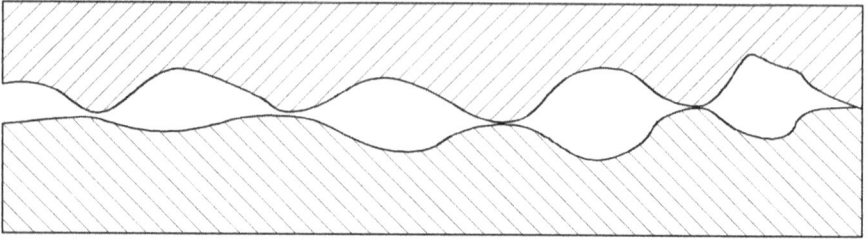

FIGURE 2.10 Magnified view of the interface between the cutting tool and the machined workpiece surface.

To look at the penetration of the cutting fluid inthe chip tool interface, Naerheim et al. (1986) examined the surfaces of chips while cutting 4340 steel bars and 4130 steel tubing using CCl_4 as a model cutting fluid. They did the Scanning Auger analysis of fracture surfaces of chips and they have seen the evidence and explained the penetration as the capillary action of fissures that provide reactive surfaces and fast propagation paths for the cutting fluid and vapor.

2.3 COOLING MECHANISM

A major function of a cutting fluid is cooling the workpiece and the cutting tool. During the 19th century, it was assumed that metal working fluid improved the cutting performance by its cooling properties only. Hence, they were called coolants. Also, the fact that most of the tool wear mechanisms are temperature dependent, cooling the chip cutting tool interface helps to retain the original properties of the cutting tool, thereby prolonging the tool life. However, a reduction in the temperature of the workpiece may, under certain conditions, increase the shear flow stress of the workpiece, thereby decreasing tool life. However, it has been proven through a large number of investigations that cooling, in fact, is one of the major factors in improving the cutting performance; particularly when the cutting speeds are very high, such as in grinding or high-speed turning of ferrous alloys, then the cutting fluids are excellent coolants that would be useful.

2.4 LUBRICATING ACTION

Another important function of a metal working fluid is reduction in friction by providing excellent lubrication. Lubricating action provides the best improvement in machining performance, as it reduces the heat generated. The lubricating cutting fluid reduces the energy input to the metal cutting operation. However, if the metal working fluid will be effective, it must reach the chip cutting tool interface. In order for this to happen, the cutting fluid should reach the chip tool interface. As the chip tool contact pressure approaches a high value such as 70 MPa, it is extremely difficult for the cutting fluid to reach the chip tool interface during a high-speed operation, such as a turning tool. It has been suggested by Merchant (Armarego & Brown, 1969) that the cutting fluid will move through the asperities present in the chip tool interface by the capillary action. For applications that are carried out at low cutting speeds, the cutting fluid should have a better lubricating ability.

2.5 REHBINDER EFFECT

The Rehbinder effect has been known as a phenomenon that reduces the mechanical strength when the metal is exposed to a polar organic environment or the surface of the metal is coated with some polar organic substances. The mechanism, as explained by Rehbinder (1947), is that during the metal cutting operation, the microcracks present on the surface of the material get welded under the deformation force applied thereby increasing strength of the work material. When the cutting fluid is applied, it enters these microcracks there by preventing the welding of the material thereby reducing the strength of the material that reduces the cutting force required for deformation. This has been experimentally evaluated for many materials with a comprehensive review of the process by Kaneeda (2023).

2.6 CUTTING FLUID APPLICATION DIRECTIONS

There are three possible directions through which the cutting fluid could be applied, as shown in Figure 2.11.

2.6.1 ON THE BACK OF THE CHIP (A)

Applying the cutting fluid on the back is the most convenient, as it is readily available and the fluid nozzle can be properly oriented to this surface. No modification needs to be done to the cutting tool. However, the fluid will be more effective when it penetrates the chip and rake crevice or work and flank side of the tool to remove the heat from the

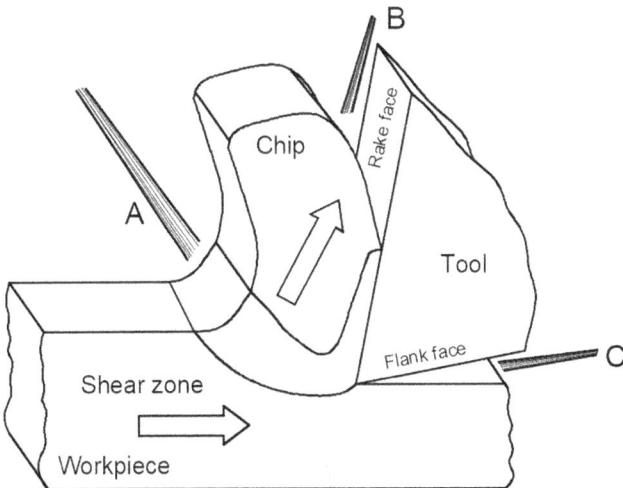

FIGURE 2.11   The directions along which cutting fluid could be applied in metal cutting.

cutting tool and reduce friction. Applied at the back of the chip does not guarantee that the fluid will be contacting the cutting tool at the appropriate location.

2.6.2 ALONG THE RAKE FACE BETWEEN THE CHIP AND RAKE FACE OF THE CUTTING TOOL (B)

When the cutting fluid is applied in this direction, it has to flow against the chip moving direction, and, as a result, it has the least amount of chance to reach the tool tip point. As such, it may not be able to serve any purpose in providing lubrication between the chip and rake face of the cutting tool. However, it may be able to help with the chip breaking.

2.6.3 ALONG THE CLEARANCE (FLANK) FACE BETWEEN THE FINISHED WORK SURFACE AND CLEARANCE FACE OF THE CUTTING TOOL (C)

Applying the flank face offers the best possible opportunity for the cutting fluid to enter into the chip tool interface as the fluid movement is in the same direction as that of the cutting tool.

In addition, most of the modern cutting tools are equipped with internal coolant through the tool such as turning tools and drills. These offer a better control of the cutting fluid jet for pinpoint location and ensure better cooling of the cutting tool, as shown in Figure 2.12. Though this type of tool is more expensive, it has many advantages in terms of cutting fluid application.

- The nozzle location should be organized in a such a way that it directs the fluid beam directly toward the cutting zone on the rake side, as shown in Figure 2.9. This would effectively reduce the cutting tool temperature and would help in chip control by chip breaking.
- It is also possible to direct the cutting fluid along the flank side of the tool, thereby providing a better cooling and lubricating function. This will ensure a higher tool life.

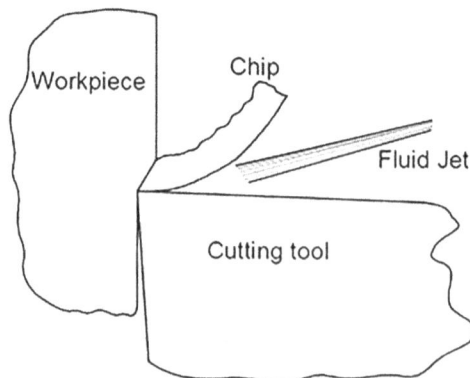

FIGURE 2.12   Modern cutting tool with coolant holes for fluid direction.

2.7 TYPES OF CUTTING FLUIDS

There is a wide variety of cutting fluids that have been developed for use in metal cutting operations.

2.7.1 WATER-BASED EMULSIONS

This is one of the largest volumes of metal working fluids used; also called water-soluble oils as the main ingredient of these fluids is water. Because of its highest heat carrying capacity (high specific heat), pure water is considered the best cutting fluid. It is also cheap and easily available. Its low viscosity makes it flow at high rates through the cutting fluid system and also penetrates the cutting zone easily. Corrosion of the work material is a major problem with water, in view of the high temperatures prevalent in the cutting zone as well as exposed to the nascent machined surface. Corrosion of the machine tool parts is also possible with pure water, since it is likely to spill during the machining operation.

A number of additives are added to water to serve these auxiliary functions. Some of the additives that are frequently added are rust inhibitors, chemicals that promote wetting characteristics, and any other additives to improve lubrication characteristics. The concentrated oil is normally diluted in water to any desired concentration, such as 30:1 to 80:1.

2.7.2 STRAIGHT MINERAL OILS

These are the pure mineral oils without any additives. The main ingredient being oil, these provides excellent lubrication. They also help with rust prevention. They may not provide much cooling. These are chemically stable and lower in cost. Because of their limited effectiveness as metal working fluids, these are used for limited applications such as light duty machining only.

2.7.3 MINERAL OILS WITH ADDITIVES (NEAT OILS)

Mineral oils are also called neat oils. For commercial applications, this is the largest variety of metal working fluids developed for different materials and processes. To cope with the requirements of different materials, a large number of additives were developed. These additives when added in appropriate amounts to the mineral oils will be able to cater to the different needs of the materials during machining operations. These cutting fluids are quite helpful for many difficult-to-machine situations.

The additives, while providing the chemical activity, improve the load carrying capacity. The additives used generally are the following:

- Fatty oils for the load carrying properties.
- EP (extreme pressure) additives used for more difficult-to-machine situations. EP additives react with the material and form low shear stress materials, such as iron sulphide that has low shear strength and will act as a

solid lubricant in that location to reduce the welding (sticky region) between the work and the tool. EP additives are basically chlorine, sulfur, or combination of both of them. The additives make the disposal of cutting fluids a major problem.

2.7.4 SYNTHETIC CUTTING FLUIDS

These are water based with some additives. These will not use mineral oils or petroleum-based oils. They use organic and inorganic compounds along with water. These are inherently lower foaming than soluble oils or semi-synthetics as they will not be using the emulsifiers. These generate less mist during the metal cutting operation. These are quite useful for the tool cutting fluid application. The sump life is more and are very clean as these reject tramp oil better.

2.8   CUTTING FLUID APPLICATION METHODS

The actual application of the cutting fluid can be in any of the following methods:

- Flooding
- HiJet application
- MQL (Mist) application

2.8.1   FLOODING

You might have seen in machine shop videos a flood of milky white fluid splashing on the cutting zone. This is termed *flooding*. A very high-volume flow of the cutting fluid is applied through direction A in Figure 2.11. The result is that the entire machining zone is flooded by the cutting fluid. The thinking is that the flow of the large volume of the cutting fluid will take away a good chunk of the heat generated in machining. The metal working fluid after completing its task will be collected and rerouted back to machining after proper treatment. Its usefulness depends upon how well it will be able to remove heat from the zone. For this purpose, sometimes smaller nozzles are used in place of a single nozzle, thereby reducing the flashing.

The problem with flooding is that the fluid may not be able to penetrate the chip-cutting tool interface in view of the large resistance present at the tool tip. To compensate, a high-pressure (up to 4 MPa) fine jet of cutting fluid is applied through the clearance face. This is termed *HiJet application* or simply *jet application*. It has been claimed that penetration in this case is higher than flooding. As a result, better performace has been found with the HiJet application of cutting fluid.

In the mist application, very small droplets of the cutting fluid are dispersed in a gas medium, generally air, and the mixture is applied at the cutting zone, through the clearance crevice. This method combines the attractive properties of gases and liquids. The advantages claimed for the mist application (Rao & Arora, 1977) include the following:

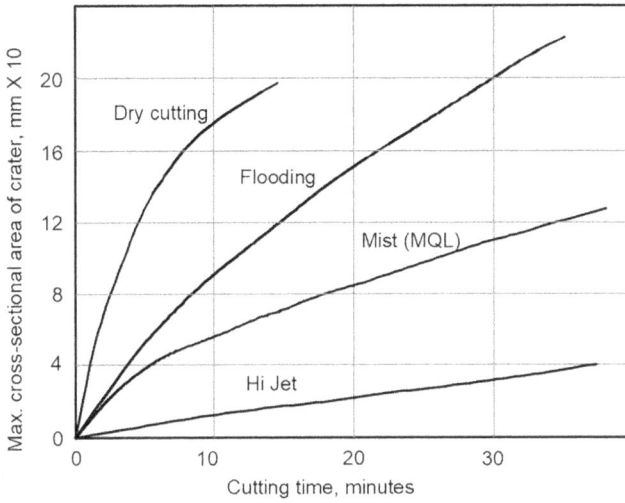

FIGURE 2.13 Comparative performances of cutting fluid application methods (Rao, 1973; Rao & Arora, 1977).

- *"Large surface to volume ratio for each drop provides the possibility of rapid vaporisation, which is an important step that must precede penetration of the chip tool interface.*
- *The small size of the particles improves the penetrating ability of the cutting fluids.*
- *The consumption of the cutting fluid is much less of the order of 300ml/ hour.*
- *There is no reclamation of the cutting fluid from the chips nor frequent cleaning of the sumps.*
- *The compressed air in the mist helps keep the chips away from the cutting zone, thus allowing the operator to follow the machining layouts easily.*
- *No spillage of the cutting fluid, thus keeping the machine, the operator, and the surroundings clean."*

It has been reported that the mist application gives generally low flank wear (Figure 2.13) and better surface finish. Its effectiveness increases at higher feed rates.

REFERENCES

Armarego, E.J.A., & Brown, R.H. (1969). *The machining of metals*. Englewood Cliffs, NJ: Prentice Hall.

Boothroyd, G., & Knight, W.A. (1989). *Fundamentals of machining and machine tools*, 2nd ed. New York: Marcel Dekker.

Cook, N. H. (1966). *Manufacturing analysis*. Reading, MA: Addison-Wesley Publishing Company.

Kaneeda, T. (2023). The lubricant applying effect: Review and considerations. *Precision Engineering*, 79, 277–290.

Naerheim, Y., Smith, T., & Lan, M.-S. (1986). Experimental investigation of cutting fluid interaction in machining. *Journal of Tribology*, 108 (3), 364–367 (10.1115/1.3261205)

Rao, P.N. (1973). Some Effects of Mist Application of Cutting Fluids in Machining Steel, M.E. dissertation submitted to BITS, Pilani.

Rao, P.N., & Arora, R.P. (1977). An evaluation of different cutting fluids and their methods of application, Proceedings of the International Conference on Production Engineering, IE (I) and CIRP, New Delhi, India, pp. v-136–v-146.

Rehbinder, P.A. (1947). New physico-chemical phenomena in the deformation and mechanical treatment of solids. *Nature*, 159, 886–867.

Shaw, M. C. (2005). *Metal cutting principles*, 2nd ed. London: Oxford University Press.

Taylor, F.W. (1907). *On the art of cutting metals*, Trans. Baltimore, MD: American Society of Mechanical Engineers.

Trent, E.M., & Wright, P.K. (2000). *Metal cutting*, 4th ed. London: Butterworth Heinemann.

3 Classification of cutting fluids and additives for cutting fluids

3.1 INTRODUCTION

Cutting fluids are primarily used to lubricate and cool the machining zone. These fluids also help in decreasing the adhesion between the rake face of the tool and the chip, this reducing the adhesion wear. Further, the cutting fluids also cause chip curl and reduce the rake contact length. Apart from these primary functions, several auxiliary functions like corrosion protection, washing away of the chips, etc. are performed by the cutting fluids.

The cooling action of the fluid prevents thermal expansion of the workpiece. This helps to prolong the life of the cutting tool and obtain better surface finish of the product. All these benefits resulted in the popular use of cutting fluids in machining. The chronological development of cutting fluids is presented in Table 3.1.

Different types of cutting fluids are available in the market: straight oils (water miscible oils (emulsions), synthetic fluids, and semi-synthetic fluids (Figure 3.1). Of these, water miscible oils are used in 80 to 90% applications worldwide. Though straight oils are not currently as common as in the past, they are still the fluids of choice for certain metal cutting applications.

3.2 TYPES OF CUTTING FLUIDS

Operational requirements are dictated mostly by the materials of the workpiece and the cutting tool involved, process to be performed, product quality, cost, and other factors. The choice of cutting fluid is also based on parameters like hardness of water, chemical constraints, filtration, etc.

Initially, mineral oils were used as cutting fluids to lubricate the machining zone. Additives like lard cutting fluids were simple oils applied with brushes to lubricate and cool the machine tool. Occasionally, lard or other animal-based fats were added to increase the lubricity. However, with the growing needs of the machining industry, more complex fluid compositions were invented and used. Currently, the cutting fluids contain various additives to increase the performance and life of the fluids. The cutting fluids can be broadly categorized as follows.

3.2.1 STRAIGHT OILS

Straight oils are petroleum or mineral oils, and do not contain any water. Several additives may be added to improve the performance of the fluids. For instance,

DOI: 10.1201/9781003328742-3

TABLE 3.1

Chronological development of cutting fluids (Wickramasinghe et al., 2021)

Time span	Driving factor	Effect of cutting fluid composition
Before 1800	Demand to machine metals	Development of first MWFs based on natural products e.g., water, animal or vegetable oil.
1800–1899	Industrialization (machine tools) Ability of Mineral oil	Replacement of natural MWF components. First investigation on the lubrication ability of the mineral oil–based MWFs.
1900–1999	Superior tool and workpiece materials Advanced machine tools Mass production	Addition of numerous chemical substances to increase the technical performance. Application of chlorinated MWFs containing boric acid and other harmful chemicals. First approach to reduce amount of mineral oil in MWFs (driven by the rising oil price).
2000–up to date	Health, safety, and environmental regulations Energy and resource efficiency	Substitution or elimination of chlorine and other harmful substances. Assessment of the sustainability. Interdisciplinary assessment of MWFs.

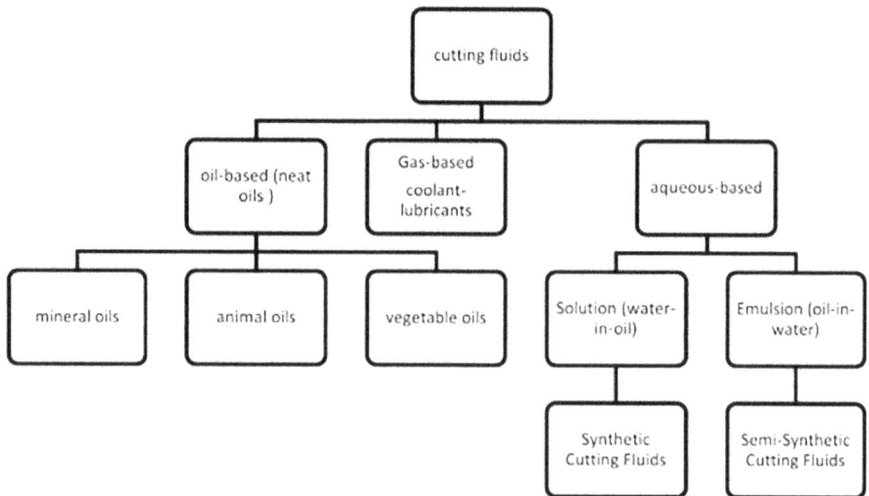

FIGURE 3.1   Classification of cutting fluids (Debnath et al., 2014).

extreme pressure (EP) additives may be added for heavy-duty machining like gear cutting. For normal machining, additives may not be required. Since the straight oils are just the oils, they provide the best lubrication; however, they do not help much with the heat removal. So, these oils are best suited for low-speed operations requiring high surface finish. The high cost of the oils is justified by the savings that are obtained in terms of the longer tool life due to better lubrication. These cutting fluids are used in various operations like severe braoching, tapping, deep hole drilling, etc. Apart from these, straight oils are populary used in machining of difficult-to-machine metals. These materials have poor thermal conductivity and the heat transfer rate is very low. Hence better lubrication is helpful. Straight oils are commonly used in minimum quantity lubricaiton (MQL) applications since lubricating the secondary shear zone is critical in those applications. Further, these fluids have better rust protection, easier maintenance, no problem of microbial contamination compared with the other types of cutting fluids. However, straight oils have poor thermal properties like thermal conductivity and specific heat. They also pose a fire hazard due to low flash and fire point temperatures. Chances of mist generation are also higher with the straight cutting fluids. Since the smoke point/flash point of these oils are low compared to the water soluble oils, these fluids should be carefully applied. Straight oils are generally used only in operationsthat produce low temperatures. Further, the oil left on the workpiece needs special effort to clean with the use of solvents.

3.2.2 Water-soluble Oils

Water-soluble oils (also referred to as emulsions) contain oil, water, and emulsifiers. They have the blended advantages of both oil (lubrication) and water (cooling). These fluids also help in preventing the corrosion of the metal surfaces. However, the performance of these fluids is greatly influenced by the hardness of the water. Hard water reduces the stability of the emulsion and reduces the cooling/lubricating abilities of the cutting fluid.

The emulsifiers (soap-like materials) help in dispersing the oil particles in water and form a stable oil-water emulsion. The oil particles cling to the workpiece and help in lurbcation in the machining. Sodium petroleum sulphonate (SPS) is popular (Figure 3.2). These emulsifiers act as surfactants. The sulphonate end of the SPS

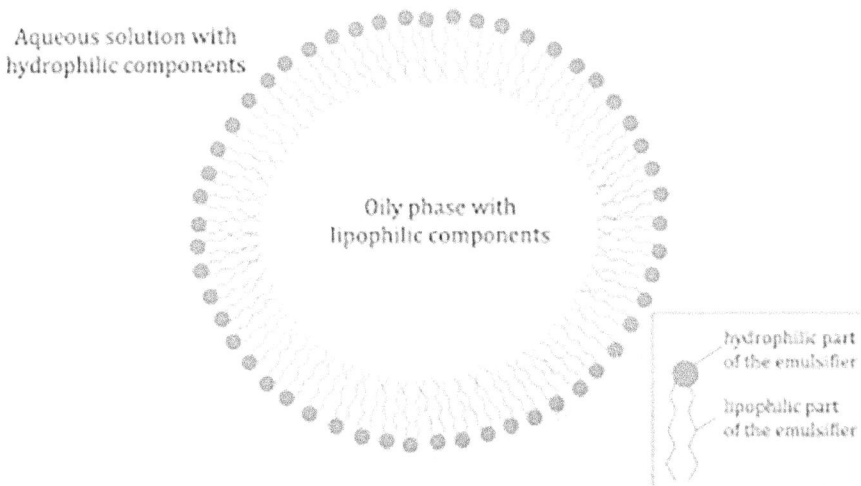

Aqueous solution with hydrophilic components

Oily phase with lipophilic components

hydrophilic part of the emulsifier

lipophilic part of the emulsifier

FIGURE 3.2   Emulsifier molecule in action (Brinksmeier et al., 2015).

molecule is polar and acts as an anionic surfactant. This end holds the water molecules, while the head of the SPS molecule has no charge and it holds the oil molecules. Thus, both oil and water are held together to make an emulsion. Due to refraction, the emulsifier particles give a milky white color to the fluid.

Typical composition of thse cutting fluids consists of about 3.5–4% of mineral oil, 0.5–1% of emulsifier, 95% of water and various additives for the remaining part (Brinksmeier et al., 2015).

3.2.3 SYNTHETIC FLUIDS

Synthetic fluids, as their name suggests, are made in the lab and do not contain petroleum or mineral oil. These fluids contain chemical lubricants like $CHCl_3$ and rust inhibitors mixed with water. Like water-soluble oils, synthetic cutting fluids are in the form of a concentrated liquid. These are then mixed with water to obtain the actual cutting fluid. Different properties of these fluids can be adjusted as required by the machining operation. which are mixed with water to form the cutting fluid. Because of their better cooling abilities, these fluids are preferred for high-speed operations. They are also desirable when the fluid needs to be clear and should not have excessive foaming characteristics.

In addition to the regular synthetic oils, heavy-duty synthetics are being used in the recent years. These fluids are useful for most machining operations. The most common chemicals added to the synthetic fluids include:

- Rust inhabitors
- Water softeners
- Soaps and wetting agents
- Biocides

Since synthetic fluids do not contain the regular oils, they have excellent microbial control and resistance to rancidity.

However, it is important to note that although synthetics are less problematic compared to the oil-based fluids, they are not completely devoid of them. Various concerns like mist formation and causing skin disorders to the workers are present even when using the synthetic fluids in machine. Some of the ingredients that are added to increase the lurbicity of the fluid may emulsify the tramp oils, or leave residues. Also, these fluids can react with other fluids like the lubricants of the machine tool and get contaminated. The synthetic fluids needs to be carefully monitored for good performance.

3.2.4 SEMI-SYNTHETIC FLUIDS

Semi-synthetic fluids are combinations of emulsions and synthetic fluids. About 2–30% of the mineral oil is dispersed in a water-dilatable concentrate. The remaining portion of the concentrate contains emulsifiers and water. Several other ingredients like wetting agents, corrosion inhibitors, and biocides are added. Semi-synthetic fluid concentrate already contains water; the concentrate need not be diluted again. The emulsification of oil and water happens during the manufacturing process of the fluid. This is a unique characteristic. Due to the high emulsifier content, semi-synthetics tend have small-sized oil globular suspensions. This reduces the amount of light refracted by the fluid and keeps the fluid more clear compared to the emulsions.

Semi-synthetic fluids commonly display a range of translucency levels, from almost transparent with a slight haze to completely opaque. Moreover, most of the semi-synthetics are sensitive to heat. In these fluids, the oil molecules tend to gather around the cutting tool, providing an increased level of lubrication. As the fluid cools, the molecules redistribute throughout the solution once again.

Similar to the synthetics, semi-synthetics perform well in different machining applications and are very easy to maintain compared to soluble oils. These fluids provide good lubrication in various applications ranging from moderate to heavy machining. Due to the better cooling and wetting properties compared to the soluble cutting fluids, these fluids can be used in high-speed operations and higher feed rates. Semi-synthetic fluids have less viscosity, so they do not settle down easily. Semi-synthetics control rancidity, microbial contamination, smoke generation, and mist. Misting is less due to the low content of oil. These fluids also have a longer life.

However, similar to the emulsions, water hardness is a major concern in these fluids. Hard water may cause sum deposits in the fluid and increase the maintenance. Also, since various chemicals are added, these chemicals may cause excessive foaming.

3.2.5 GASEOUS COOLANTS

Apart from using liquids as coolants in machining, gases at sub-zero (cryogenic) temperatures are also used. Greater wear resistance is exhibited by tools at low temperatures (Zindani and Kumar, 2020). A protective layer is formed on tools that reduces flank wear in cryogenic cooling (Figure 3.3).

Also, the machined components have better wear resistance and fatigue life. Different gases like CO_2 and liquid N_2 are supplied at high pressures and at cryogenic temperatures. A typical cryogenic machining setup is shown in Figure 3.4. It is also reported that cryogenic machining produces cleaner machined surfaces than dry or wet machining (Bordin et al., 2017). However, the cost of equipment, energy

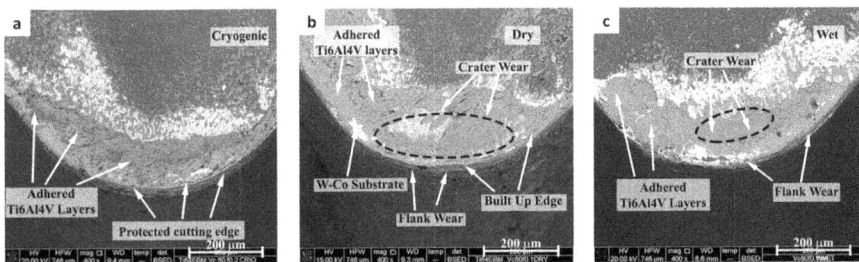

FIGURE 3.3   SEM images of tool after 15 minutes of machining: (a) cryogenic cooling, (b) dry cutting, (c) wet machining (Bordin et al., 2017).

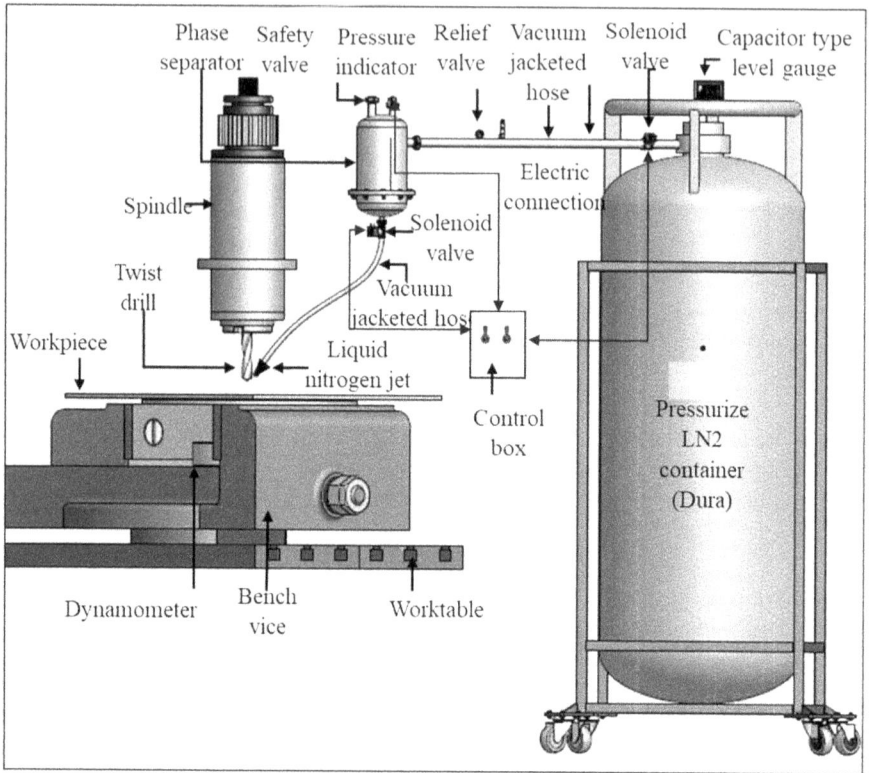

FIGURE 3.4 Schematic of cryogenic machining setup (Khanna et al., 2021).

consumption, and the cost of the fluid is high, thus limiting the commercial viability of these fluids.

Though several alternatives exist, water-soluble oils are most commonly used due to commercial viability and performance.

3.3 VEGETABLE-BASED CUTTING FLUIDS

Due to the environmental concerns, the mineral oils in the straight oil or water-miscible oils are sometimes replaced with vegetable-based oils. These oils are biodegradable and nontoxic. This helps in safe and effective disposal of the cutting fluids. Further, these oils have triglycerides and long structures that help in reducing friction (Syahir et al., 2017). Though the vegetables have many advantages, they are not commercially popular due to poor oxidation stability and poor ability to withstand higher temperatures. Apart from the functional aspects, the cost of vegetable oils is sometimes prohibitive for their commercial application (Figure 3.5.). Table 3.2 compares the properties of different vegetable oils with mineral oil.

In turning operation, about 70% of the generated heat is carried away by the chip. The remaining heat is contained in the tool and the workpiece. This can be carried away by using a cutting fluid. However, if a cutting fluid is not used, this

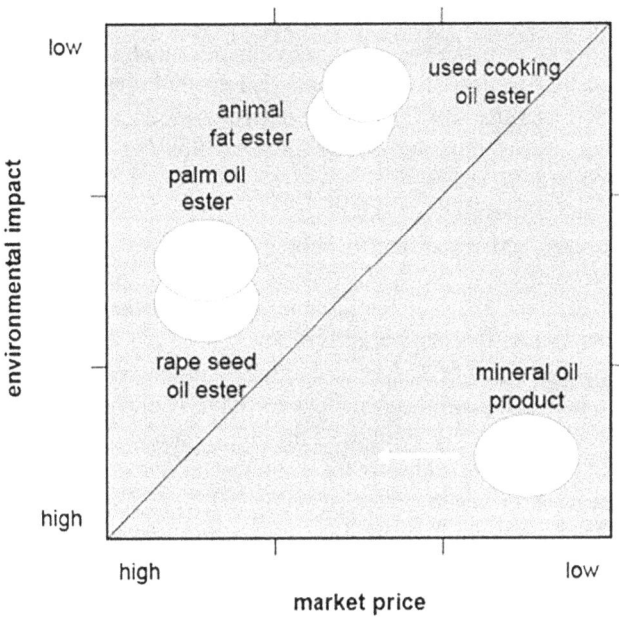

FIGURE 3.5 Market price/environmental impact portfolio (Wickramasinghe et al., 2020).

TABLE 3.2

Comparison of the properties of vegetable oils and mineral oil (Wickramasinghe et al., 2020)

Performance	Canola oil (vegetable oil)	TMPTO (polyol ester)	Sat/complex (synthetic ester)	PAG (petroleum synthesized)	Mineral oil (petroleum)
Biodegradability	Excellent	Better	Better	Good	Poor
Toxicity	Low	Low	Low	Low	High
Oxidative stability	Poor	Moderate	Better	Good	Better
Lubrication	Excellent	Better	Better	Better	Good
Heat Exchange	Poor	Good	Good	Good	Good
Viscosity Index	Very good	Better	Better	Better	Moderate
Hydrolytic stability	Poor	Moderate	Good	Good	Very good
Thermal stability	Moderate	Good	Better	Good	Good
Seal compatibility	Moderate	Moderate	Moderate	Good	Better
Relative cost with respect to mineral oil	2	2	6–8	4	1

heat remains in the tool and workpiece resulting in increased tool and workpiece temperatures. Such high temperatures not only cause dimensional deviations in the workpiece and thermal failure of the cutting tools, but also damages the workpiece by inducing residual stresses and micro-cracks on the workpiece. Increased temperatures of the workpiece make it more prone to oxidation and corrosion (Chiou et al, 2003; Ko et al., 1999; Sukaylo et al., 2005). Further, high machining temperatures make the tool more vulnerable and damage the cutting tool. In some cases, built-up edge (BUE) is formed on the tool tip (Chou, 2003; Ko et al., 1999). BUE increases the surface roughness of the product and also damages the tool. The above factors have led to the investigations to develop cutting fluids having effective cooling action.

Sales et al. (2002) deviced a method to estimate the cooling abilities of cutting fluids. The authors used a tool-thermocouple technique to measure the chip-tool interface temperature under the application of various cutting fluids in the turning operation. It was noted that the temperature dropped from 300°C to room temperature for a heated AISI 8640 workpiece. The workpiece was held on a lathe and rotated at a constant speed of 150 rpm. Cutting fluid was applied at a fixed point. Using the measured temperatures, the convective heat transfer coefficient was estimated to quantify the cooling abilities of the cutting fluids. The transfer coefficient was determined to estimate and rank the cooling abilities of the cutting fluids.

Vieira et al. (2001) tested water miscible oil, semi-synthetic, and synthetic cutting fluids in the face milling of AISI 8640 steel with coated cement carbide tools. As a reference, dry machining was also carried out. Tool life, power consumption, and surface roughness were measured in the experiments. Cutting temperatures during the turning of AISI 1020 steel were measured using the tool-thermocouple technique. It was reported that while the highest cutting temperatures were reported in dry machining, synthetic fluids, and semi-synthetic fluids gave lesser temperatures.

Cambiella et al. (2007) performed EP tests using ASTM D 2783 method. Different emulsion-based cutting fluids were used in machining operations. Different emulsifiers, anionic, non-ionic, and cationic surfactants, were used in different concentrations in the machining experiments. It was reported that the performance of the cutting fluids varied based on the type and concentration of the emulsifier.

With increasing environmental awareness, vegetable oils are being used in place of the petroleum-based oils in the cutting fluids (Figure 3.6).

De Chiffre and Belluco (2001) studied the performance of cutting fluids under different machining operations: turning, drilling, reaming, and tapping. Different parameters like tool life, cutting focus, surface integrity, and dimensional accuracy were studied. Austenitic stainless steel was machined under the application of water miscible and straight oils. It was reported that vegetable oils and esters were best in all operations whether they were water-based fluids or straight oils. However, the performance of the straight oils varied with the workpiece material.

Mould et al. (1972) tested a different organochlorine compounds as cutting fluids in tapping. 15 mg Cl/100 g was dispersed in liquid paraffin. The performance of the different fluids in machining was mapped with the performance of the same fluids in a four-ball test. A strong correlation was observed in the results. It was also found that the concentration of the blends improved the performance with respect to the cutting forces.

In many cases, cutting fluids are looked upon as additional expenditure to the tune of 15–25% of the total cost of production (Jen et al., 2002). However, Motta and Machado (1995) commented that the savings that are obtained in the machining

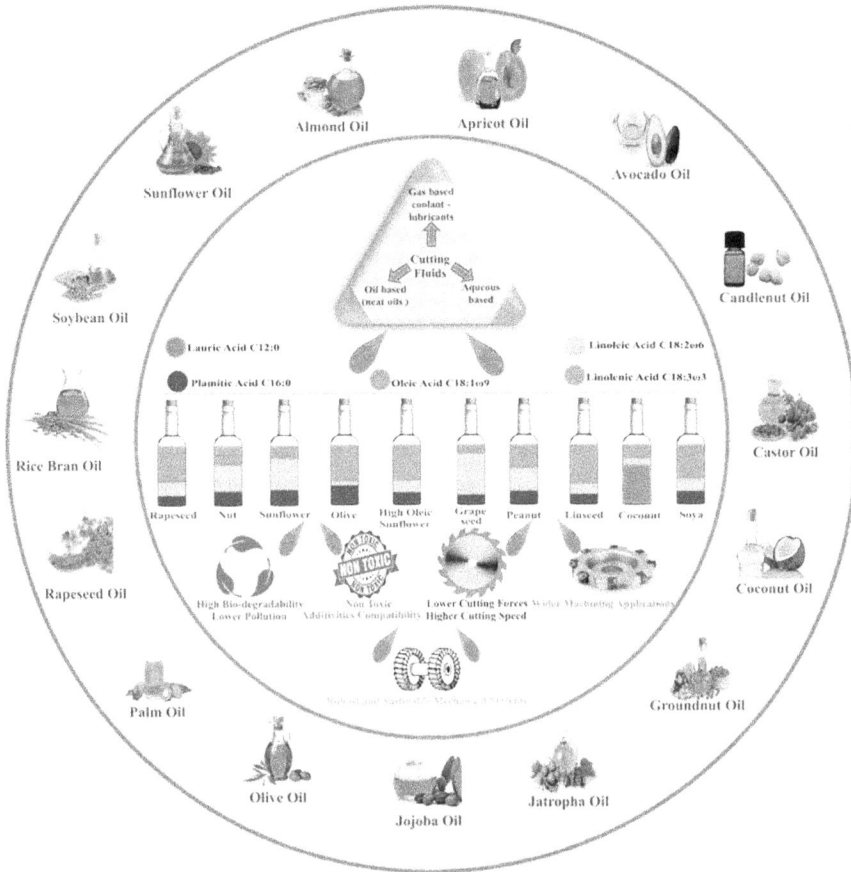

FIGURE 3.6 Vegetable oils in machining (Sankaranarayanan et al., 2021).

through longer tool life, energy savings justify the cost of the cutting fluids. Minke (1999) pointed out that cutting fluids are critical in improving and maintaining the quality of the product. It was pointed out that straight oils have higher lubrication while emulsions have better cooling abilities.

Belluco and De Chiffre (2002) studied the performance of novel formulations of vegetable-based cutting fluids on dimensional accuracy, micro-hardness, and surface finish. AISI 316 L steel was reamed and tapped under the application of these fluids. A regular cutting fluid was used as a reference for all experiments. It was found that the cutting fluids played a critical role in the measurement of different parameters. It was found that vegetable oils had comparable or even better performance compared to mineral oils.

Chang et al. (2022) compared the performance of emulsified palm oil/coconut oil, and virgin palm oil/coconut oil while machining titanium alloys. It was found that virgin oils performed better than emulsified oils (Figure 3.7). This was attributed to the lower coefficient of friction for the virgin oils.

(a)

(b)

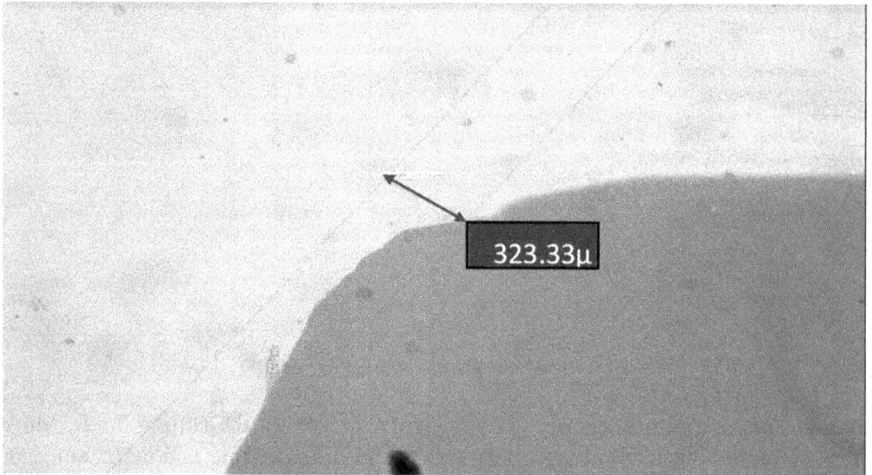

FIGURE 3.7   Flank wear of cutting tool using: (a) virgin palm oil, (b) emulsified palm oil (Chang et al., 2022).

In a later work, Belluco and De Chiffre (2002) compared the performance of the six different vegetable-based cutting fluids in drilling AISI 316 L austenitic stainless steel. HSS-Co tools were used in the machining. Tool life, tool wear, cutting forces, and chip formation were studied in the experiments. The cutting tools were tested until the tools completely broke. A regular emulsion was used as a benchmark fluid. Through the measurement of the different parameters, it was reported that the cutting fluids performed better than the benchmark fluid. In one of the cases, a 177% increase in tool life and a 7% reduction in thrust force were reported.

Baradie (1996) reported an increase in tool life and improvement in surface texture with the application of cutting fluids. It was reported that dimensional accuracy and energy consumption improved with the application of cutting fluids. The ratio of oil to water of 1:40 or 1:50 was found to be suitable for operations that require a lighter fluid with more cooling action. Water-soluble fluids provided both lubrication and cooling that are required in high speed machining.

Rakić and Rakić (2002a) studied the influence of the application of cutting fluids while machining on 30 lathes. The role of cutting fluids in the failure of machine tools was studied. It was reported that water-based fluids influenced the tribological processes, wear, and failures of the machine tools. Wang and Kou (1997) studied the application of cutting fluid in grinding. The results revealed that water had a better cooling effect than oil. Further, cutting fluid had better cooling abilities at lower workpiece speed, higher grinding depth, and higher wheel speed. Avila and Abrão (2001) evaluated the performance of different cutting fluids: two emulsions and one synthetic fluid. The performance of the fluids was compared with dry cutting in the machining of hardened AISI 4340 steel. The tool life, surface finish, tool wear mechanisms, and chip form were measured. It was observed that the fluid without the mineral oil gave the best results, followed by dry machining, synthetic fluid, and fluid with mineral oil in the order. Both synthetic fluid– and the mineral oil–based emulsion gave very similar results.

Cakir et al. (2004) compared the application of gaseous cutting fluid and dry and wet machining in the machining of AISI 1040 steel. Carbon dioxide, oxygen, and nitrogen were chosen for direct injection into the cutting zone. For wet machining, 5% emulsion cutting fluid was used. Dry machining produced the highest cutting forces. Though gas application produced lesser cutting force compared to wet machining, surface roughness was similar. Tsao (2000) investigated the influence of sulfurous boric acid ester on milling. It was found that tool wear decreased with the addition of the ester.

Bianchi et al. (2004) studied the application of two different cutting fluids: a 5% emulsion and straight oil in grinding. SEM observation of the machined surfaces showed that straight oil is better than emulsion. However, the use of straight oils was discouraged due to the potential effect on the workers' health. Alves and Oliveira (2006) applied a cutting fluid based on sulphonate vegetable oil. The cutting fluid showed better performance compared to the mineral oils based on the observed surface finish. Also, the formulated fluids were more environment-friendly compared to the mineral oil. Tsao (2007) studied the performance of a cutting fluid containing sulfurous boric acid ester in the milling aluminum alloy. Cutting forces and tool wear were measured. It was reported that the cutting fluid helped in decreasing the tool wear by 12.5% and forces decreased by 10%.

Chargin (1998) studied the application of cutting fluids in the turning of single crystal silicon using a single-point diamond turning. The effect of pH of the water used in the cutting fluid on machining was studied. It was found that lower pH fluids led to increased tool wear while higher pH helped in reducing tool wear and surface roughness. Negative rake tools reduced the surface roughness but tool wear increased compared to zero rake cutting tools.

TABLE 3.3

Metal working fluid properties comparison (adapted from Wickramasinghe et al., 2020)

Property	Straight oils	Emulsions (water-based cutting fluids)	Semi-synthetic	Synthetic
Appearance	Oily	Milky	Translucent	Transparent
Lubricity	Excellent	Good	Good	Poor
Corrosion control	Excellent	Poor	Good	Good
Cooling	Low	Good	Good	Excellent
Fire hazard	High	Low	Low	Low
Microbial control	Excellent	Poor	Good	Excellent

Rakić and Rakić (2002b) conducted tests on 36 milling machines for four periods of 2,000 working hours each. Different types of applications of fluids were studied and it was found that a centralized system helped in longer tribo-mechanical system life compared to individual systems. Haan et al. (1997) applied cutting fluid in drilling of aluminum alloys and gray cast iron. Different cutting speeds, feed, cutting fluid condition, temperatures, etc. were considered. It was reported that the cutting fluid has a significant effect on the surface finish of the drilled hole.

Fusse et al. (2004) compared the performance of vegetable oils and straight oils as cutting fluids. Straight oil helped to reduce the cutting force due to better lubrication, but the vegetable-based cutting fluid was better for removing the heat from the machining zone. Surface finish was better with the cutting fluids compared to the straight oils. Mendes et al. (2006) studied the performance of cutting fluids applied as mist in two different machining operations: drilling and turning of different aluminum alloys. The cutting fluid in turning contained EP additives. It was reported that a higher flow rate of the mist reduced the feed forces but increased the torque and power consumption in drilling. The flow rate did not influence the surface finish. The cutting fluids containing EP additives helped in reducing cutting forces and surface roughness.

It is important to note that different types of cutting fluids are suited for different applications. Table 3.3 compares the properties of the fluids in different areas. Fluids are selected based on the application and required properties.

3.4 ADDITIVES IN CUTTING FLUIDS

Though the regular formulations of cutting fluids are sufficient for general applications, some heavy applications that involve high cutting forces and pressures will need cutting fluids that have better load-bearing capacities. Similarly, since the regular cutting fluids are vulnerable to microbial contaminations, adding a biocide will help to control the growth of bacteria. In this way, various additives are added to improve the performance of cutting fluids. These additives are chosen based on the application. The important additives used are as follows.

3.4.1 BIOCIDES

Microbial contamination is a major factor that affects the performance of the cutting fluids. In order to mitigate the negative effects, biocides that release formaldehyde are commonly used (Brinksmeier et al., 2015). Other common biocides include phenol derivatives and isothaizolinones (Wickramasinghe et al., 2020). These biocides delay the growth of microbial colonies and help to control the contamination. Though the use of biocides is strongly discouraged by various environmental agencies, they cannot be completely avoided. Though other techniques like exposing the fluid to UV or gamma radiation are available, they are not very popularly used. It is interesting to note that microbial contamination cannot be completely avoided with the use of biocides (Trafny et al., 2015).

3.4.2 CORROSION INHIBITORS

Corrosion inhibitors help to decrease the corrosion rate of metallic surfaces. Cutting fluids contain water in most of the cases. Since cutting fluids are constantly used in the machine tools and are often used in the machining of steels, it is important that the cutting fluids do not corrode the surfaces. In the early days, inorganic corrosion inhibitors containing nitrates, chromates, zinc salt, etc. were used. But due to the ecological issues, they are now being replaced by organic compounds like alkanolamine, caborxylic acid, etc. (Li et al., 2022). The corrosion inhibitors form a layer on the metallic surface and prevent corrosion. Cutting fluids contain a combination of various corrosion inhibitors based on the application.

3.4.3 EXTREME PRESSURE ADDITIVES

Extreme pressure (EP) additives are added to the cutting fluids when used in operations that involve high pressures and cutting forces. EP additives can function at higher temperatures than the regular cutting fluids. EP additives are generally sulphur/chlorine/phosphorus-based organic compounds like chlorineparaffine, sulphurous ester, phosphoric acid ester, and polysulphide (Wickramasinghe et al, 2020) that react with the chips/workpiece to form sulfides/chlorides/phosphates. The EP additives are selected based on the temperature of the application (Azarhoushang et al., 2021).

These newly formed compounds act as solid lubricants and reduce the friction.

However, the EP additives are non-biodegradable and their disposal is difficult. Hence, the use of the EP additives is discouraged.

3.4.4 DEFOAMERS

Constant circulation of the cutting fluid leads to the formation of foam. Cutting fluids contain surfactants that aid in the formation of foam. Surfactants reduce the surface tension, thus increasing the surface area and producing foam. Foam affects the lubricating and cooling effects of the cutting fluid. Further, foam contains pockets of air, which can cause oxidation. This reduces the quality of the cutting fluid.

Also, when the coolant tank contains foam, the fluid can overflow causing oil spills. This leads to changes in the concentration. All major actions of the fluid including rust resistance, flow, etc. are affected by the foam.

In order to stop the foam, silicone polymers or tributyl phosphate-based defoamers/anti-foaming agents are added to the cutting fluids. The silicone-based defoamers are stable even at higher temperatures and high pressures.

3.4.5 WETTING AGENTS

Besides cooling and lubricating, cutting fluids also help in removing the chips. This is possible only if the cutting fluid is able to wet the surface of the chips. One way of estimating the wetting ability of a fluid is to calculate the contact angle of a drop on the metal surface (Figure 3.8). In order to improve the wetting nature of the

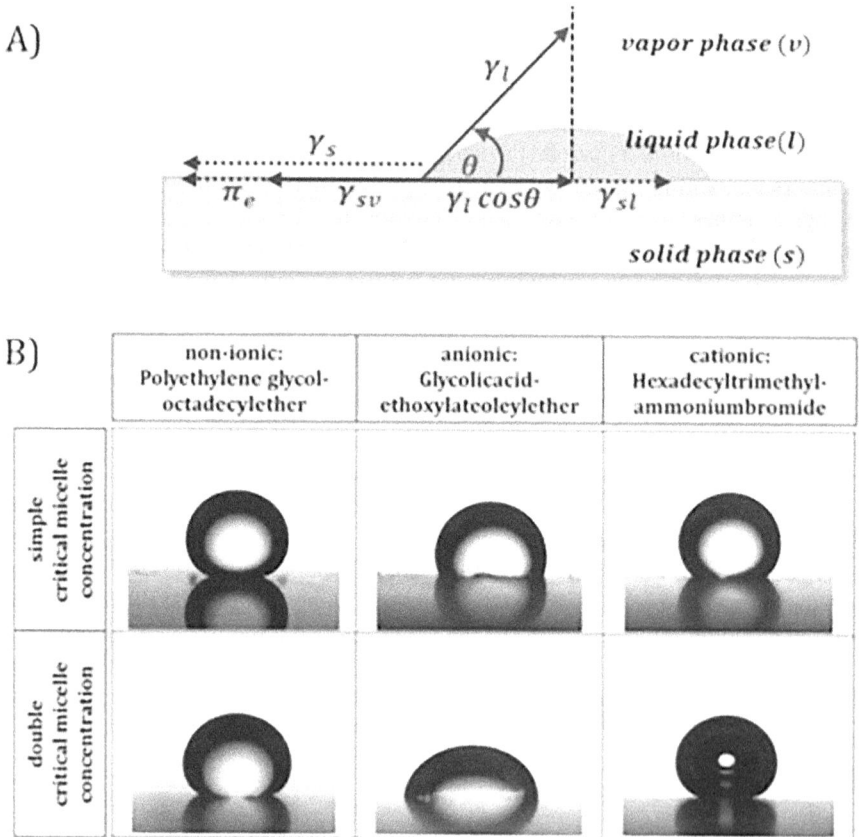

FIGURE 3.8 Contact angle of various droplets on machined steel surface (Brinksmeier et al., 2015).

fluids, non-ionic surfactants are added to the cutting fluids as wetting agents. The wetting agents reduce the surface tension of the fluids and help in wetting the chips and other surfaces. These are mainly used in synthetic fluids (Figure 3.8).

In the case of water-miscible cutting fluids, the wetting ability depends on the hardness of water, apart from factors like mixing of tramp oil and metal particles. Tramp oil and metal particles can be controlled through filtration. However, hard water contains cations that reduce the wetting ability of the fluids. So, water treatment is an effective method to improve the wetting ability of metal working fluids.

3.5 CONCLUSION

Different blends of cutting fluids are available and can be chosen based on the application. Straight oils, water-miscible oils, synthetics, semi-synthetics, and gaseous fluids are available. In recent times, vegetable oils are also being used to replace the mineral oils. Different additives are added to the cutting fluids to increase their performance.

REFERENCES

Alves, S. M., & de Oliveira, J. F. G. (2006). Development of new cutting fluid for grinding process adjusting mechanical performance and environmental impact. *Journal of Materials Processing Technology*, 179(1-3), 185–189.

Avila, R. F., & Abrao, A. M. (2001). The effect of cutting fluids on the machining of hardened AISI 4340 steel. *Journal of Materials Processing Technology*, 119(1–3), 21–26.

Azarhoushang, B., Marinescu, I. D., Rowe, W. B., Dimitrov, B., & Ohmori, H. (2021). *Tribology and fundamentals of abrasive machining processes*. USA: William Andrew.

Belluco, W., & De Chiffre, L. (2002). Surface integrity and part accuracy in reaming and tapping stainless steel with new vegetable based cutting oils. *Tribology International*, 35(12), 865–870.

Belluco, W., & De Chiffre, L. (2004). Performance evaluation of vegetable-based oils in drilling austenitic stainless steel. *Journal of Materials Processing Technology*, 148(2), 171–176.

Bianchi, E. C., Franzo, C. G., Aguiar, P. R. D., & Catai, R. E. (2004). Analysis of the influence of infeed rate and cutting fluid on cylindrical grinding processes using a conventional wheel. *Materials Research*, 7, 385–392.

Bordin, A., Sartori, S., Bruschi, S., & Ghiotti, A. (2017). Experimental investigation on the feasibility of dry and cryogenic machining as sustainable strategies when turning Ti6Al4V produced by Additive Manufacturing. *Journal of Cleaner Production*, 142, 4142–4151.

Brinksmeier, E., Meyer, D., Huesmann-Cordes, A. G., & Herrmann, C. (2015). Metalworking fluids—mechanisms and performance. *CIRP Annals*, 64(2), 605–628.

Çakır, O., Kıyak, M., & Altan, E. (2004). Comparison of gases applications to wet and dry cuttings in turning. *Journal of Materials Processing Technology*, 153, 35–41.

Cambiella, A., Benito, J. M., Pazos, C., Coca, J., Hernández, A., & Fernández, J. E. (2007). Formulation of emulsifiable cutting fluids and extreme pressure behaviour. *Journal of Materials Processing Technology*, 184(1–3), 139–145.

Chang, K. S., Olugu, E. U., Yeap, S. P., Abdelrhman, A. M., & Aja, O. C. (2022). Virgin and emulsified vegetable oil on the turning of titanium alloy. *Materials Today: Proceedings*, 48, 888–894.

Chargin, D. (1998). *Cutting fluid study for single crystal silicon* (No. UCRL-ID-130754). Livermore, CA: Lawrence Livermore National Lab. (LLNL).

Chiou, R. Y., Chen, J. S., Lu, L., & North, M. T. (2003, January). The effect of an embedded heat pipe in a cutting tool on temperature and wear. In *ASME International Mechanical Engineering Congress and Exposition* (Vol. 37181, pp. 369–376).

Chou, Y. K. (2003). Hard turning of M50 steel with different microstructures in continuous and intermittent cutting. *Wear*, 255(7–12), 1388–1394.

De Chiffre, L., & Belluco, W. (2001). Correlation of cutting fluid performance in different machining operations. In *5th Convegno Associazione Italiana di Tecnologia Meccanica*.

Debnath, S., Reddy, M. M., & Yi, Q. S. (2014). Environmental friendly cutting fluids and cooling techniques in machining: a review. *Journal of Cleaner Production*, 83, 33–47.

EI-Baradie, M. A. (1996). Cutting fluids: part 1 characterization. *Journal of Materials Processing Technology*, 56, 786–797.

Fusse, R. Y., França, T. V., Catai, R. E., Silva, L. R. D., Aguiar, P. R. D., & Bianchi, E. C. (2004). Analysis of the cutting fluid influence on the deep grinding process with a CBN grinding wheel. *Materials Research*, 7, 451–457.

Haan, D. M., Batzer, S. A., Olson, W. W., & Sutherland, J. W. (1997). An experimental study of cutting fluid effects in drilling. *Journal of Materials Processing Technology*, 71(2), 305–313.

Jen, T. C., Gutierrez, G., Eapen, S., Barber, G., Zhao, H., Szuba, P. S., ... & Manjunathaiah, J. (2002). Investigation of heat pipe cooling in drilling applications.: part I: preliminary numerical analysis and verification. *International Journal of Machine Tools and Manufacture*, 42(5), 643–652.

Katna, R., Suhaib, M., & Agrawal, N. (2020). Nonedible vegetable oil-based cutting fluids for machining processes–a review. *Materials and Manufacturing Processes*, 35(1), 1–32.

Khanna, N., Agrawal, C., Pimenov, D. Y., Singla, A. K., Machado, A. R., da Silva, L. R. R., ... & Krolczyk, G. M. (2021). Review on design and development of cryogenic machining setups for heat resistant alloys and composites. *Journal of Manufacturing Processes*, 68, 398–422.

Ko, T. J., Kim, H. S., & Chung, B. G. (1999). Air–oil cooling method for turning of hardened material. *The International Journal of Advanced Manufacturing Technology*, 15, 470–477.

Li, H., Zhang, Y., Li, C. et al. (2022). Cutting fluid corrosion inhibitors from inorganic to organic: Progress and applications. *Korean J. Chem. Eng*, 39, 1107–1134. 10.1007/s11814-021-1057-0

Mendes, O. C., Avila, R. F., Abrao, A. M., Reis, P., & Paulo Davim, J. (2006). The performance of cutting fluids when machining aluminium alloys. *Industrial Lubrication and Tribology*, 58(5), 260–268.

Minke, E. (1999). Contribution to the role of coolants on grinding process and work results. *Technical Papers-Society of Manufacturing Engineers-All Series*.

Motta, M. F., & Machado, A. R. (1995). Cutting fluids: types, functions, selection, application methods and maintenance. *Machining and Metals Journal (Revista Máquinas e Metais)*, 44–56.

Mould, R. W., Silver, H. B., & Syrett, R. J. (1972). Investigations of the activity of cutting oil additives II. organochlorine containing compounds. *Wear*, 22(2), 269–286.

Rakić, R., & Rakić, Z. (2002a). The influence of the metal working fluids on machine tool failures. *Wear*, 252(5–6), 438–444.

Rakić, R., & Rakić, Z. (2002b). Tribological aspects of the choice of metalworking fluid in cutting processes. *Journal of materials processing technology*, 124(1–2), 25–31.

Sales, W. F., Guimaraes, G., Machado, A. R., & Ezugwu, E. O. (2002). Cooling ability of cutting fluids and measurement of the chip-tool interface temperatures. *Industrial Lubrication and Tribology*, 54(2), 57–68.

Sankaranarayanan, R., & Krolczyk, G. M. (2021). A comprehensive review on research developments of vegetable-oil based cutting fluids for sustainable machining challenges. *Journal of Manufacturing Processes*, 67, 286–313.

Shaw, M. C. (2004). *Metal Cutting Principles*, 2nd ed. USA: Oxford University Press.

Sukaylo, V., Kaldos, A., Pieper, H. J., Bana, V., & Sobczyk, M. (2005). Numerical simulation of thermally induced workpiece deformation in turning when using various cutting fluid applications. *Journal of materials processing technology*, 167(2–3), 408–414.

Syahir, A. Z., Zulkifli, N. W. M., Masjuki, H. H., Kalam, M. A., Alabdulkarem, A., Gulzar, M., … & Harith, M. H. (2017). A review on bio-based lubricants and their applications. *Journal of Cleaner Production*, 168, 997–1016.

Trafny, E. A., Lewandowski, R., Kozłowska, K., Zawistowska-Marciniak, I., & Stępińska, M. (2015). Microbial contamination and biofilms on machines of metal industry using metalworking fluids with or without biocides. *International Biodeterioration & Biodegradation*, 99, 31–38.

Tsao, C. C. (2000). Study on tool life and surface roughness in milling aluminum alloys using sulfurous boric acid ester cutting fluids by Taguchi method. In *Proceedings of the 17th Conference CSME* (Vol. 4, pp. 425–428).

Tsao, C. C. (2007). An experiment study of hard coating and cutting fluid effect in milling aluminum alloy. *The International Journal of Advanced Manufacturing Technology*, 32(9–10), 885–891.

Vieira, J. M., Machado, A. R., & Ezugwu, E. O. (2001). Performance of cutting fluids during face milling of steels. *Journal of Materials Processing Technology*, 116(2–3), 244–251.

Wang, S. B., & Kou, H. S. (1997). Cooling effectiveness of cutting fluid in creep feed grinding. *International Communications in heat and mass Transfer*, 24(6), 771–783.

Wickramasinghe, K. C., Sasahara, H., Abd Rahim, E., & Perera, G. I. P. (2020). Green Metalworking Fluids for sustainable machining applications: A review. *Journal of Cleaner Production*, 257, 120552.

Wickramasinghe, K. C., Sasahara, H., Abd Rahim, E., & Perera, G. I. P. (2021). Recent advances on high performance machining of aerospace materials and composites using vegetable oil-based metal working fluids. *Journal of Cleaner Production*, 310, 127459.

Zindani, D., & Kumar, K. (2020). A brief review on cryogenics in machining process. *SN Applied Sciences*, 2(6), 1107.

4 Laboratory evaluation of cutting fluids

4.1 INTRODUCTION

The properties of cutting fluids play a major role in their performance. For instance, the thermal conductivity of the cutting fluids is important for the cooling capability of the fluid, while the kinematic viscosity is important for the lubrication. Apart from that, it is important that the cutting fluids are safe for the workers to work with. In order to assess the performance and usability of the cutting fluids, cutting fluids are often characterized through a series of laboratory tests. These tests use the standard testing methods like ASTM to measure the different properties of the cutting fluids like stability, viscosity, thermal conductivity, microbial contamination, etc. These tests help to not only characterize the cutting fluids, but also to formulate the cutting fluids with required properties.

4.2 BASIC PROPERTIES

Due to the various functions performed by the cutting fluids, it is important to measure different properties that help to understand the performance of the cutting fluids.

4.2.1 WATER SEPARABILITY

For a cutting fluid to act as both a coolant and a lubricant, it is important that the oil and water in the cutting fluids (water-miscible fluids) exist together. This is the basic purpose of the emulsifier in the cutting fluid. If the water separability of the fluid is high, then the cutting fluid will not be a proper mixture and will not exhibit the expected performance. ASTM D-1401 describes the testing method to assess the water separability of the cutting fluid (ASTM, 2021).

The test procedure involves mixing 40 mL of the fluid with 40 mL of distilled water. The mixture then needs to be maintained at either 54°C or 82°C in a graduated cylinder and stirred for 5 minutes; 54°C is maintained when testing fluids with less viscosity and 82°C for highly viscous fluids. After stirring, the mixture is allowed to rest and water separation should be noted every 5 minutes (Lawal et al., 2015). The results are reported in case the complete separation of the oil, emulsifier, and water does not occur after 30 minutes of testing. A separation of 3 mL or less is acceptable (Figure 4.1).

Excessive oil separation leads to a non-homogenous mixture and most of the properties like thermal conductivity, pH value, etc., are not as expected, thus reducing the performance of the cutting fluids.

 DOI: 10.1201/9781003328742-4

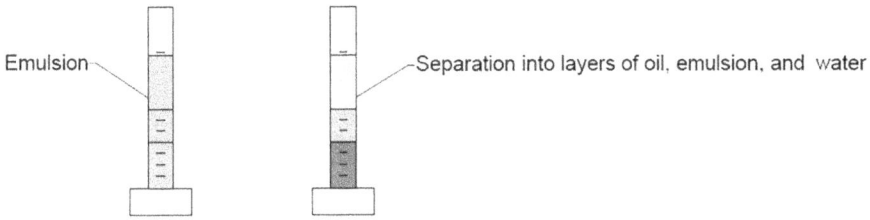

FIGURE 4.1 Illustration of water separability test.

4.2.2 KINEMATIC VISCOSITY

Kinematic viscosity is a very significant parameter to assess the effectiveness of a cutting fluid as a lubricant. Fluids with higher viscosity serve as a better lubricant in reducing friction and thus cutting forces. In the present chapter, estimation of kinematic viscosity of the cutting fluids is estimated using Redwood Viscometer-I and the viscosity of the oil-emulsifier concentrates using Redwood Viscometer-II is discussed, along with the calibration techniques. The choice of the equipment is made based on the time required for the collection of fluid (ASTM Standard D0445–04E01, 2005).

Calibration of the apparatus of the equipment is done against the kinematic viscosity of water and validated with any standard known value. Figures 4.2–4.5 show an example of calibration with distilled water and validation with SAE oil for Redwood Viscometer-1. For Redwood Viscometer-II, calibration should be done with a more viscous liquid like glycerin.

The test procedure involves taking the sample in the viscometer in either of the cases and increasing the temperature of the water bath to reach a steady state at various points. At each steady temperature, the time required for the collection of 60 mL was noted. Kinematic viscosity is calculated from the relationship,

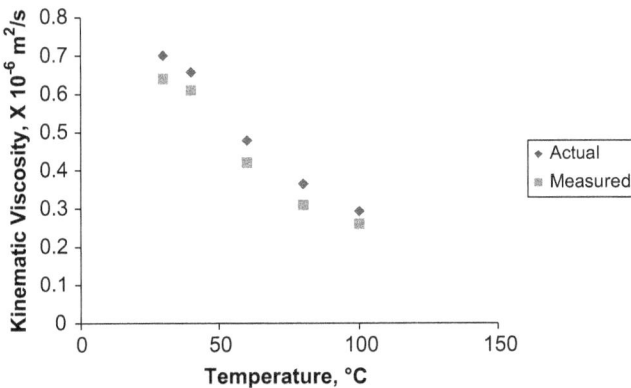

FIGURE 4.2 Comparison of actual and measured kinematic viscosity of water (Revuru, 2008).

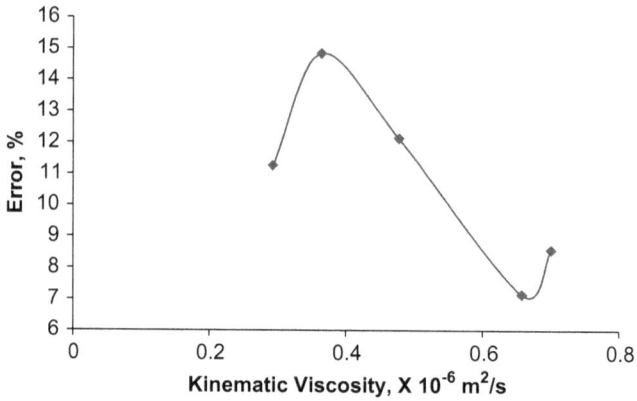

FIGURE 4.3   Error in measurement of kinematic viscosity of water (Revuru, 2008).

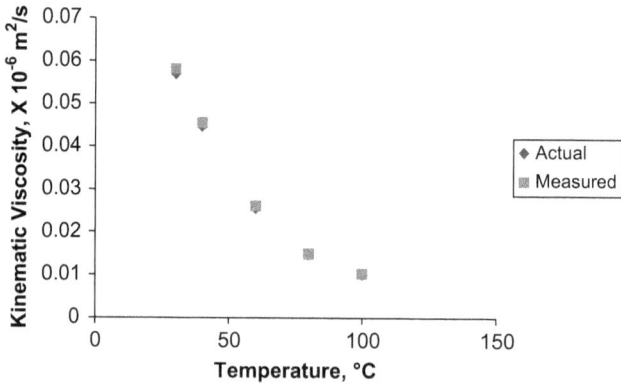

FIGURE 4.4   Comparison of actual and measured kinematic viscosity of SAE 20 oil (Revuru, 2008).

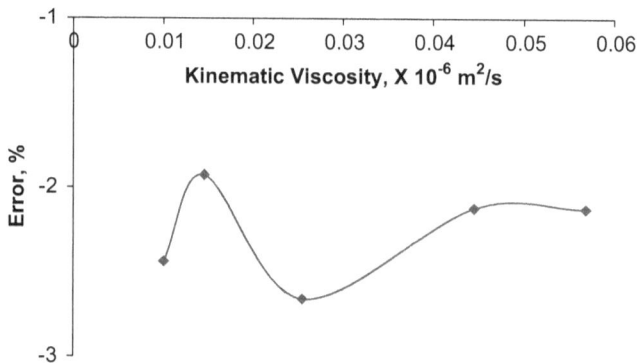

FIGURE 4.5   Error in measurement of kinematic viscosity of SAE 20 oil (Revuru, 2008).

$$\upsilon = (A^*T_r - (B/T_r)) \times 10^{-4}$$

where υ = kinematic viscosity in m^2/s

T_r = time required for collection of 60 mL, s

For, Redwood Viscometer-I,

$$\begin{aligned}
A &= 0.00264 \\
B &= 0.5, \; T_r > 100s \\
&= 1.9, \; T_r \leq 100s
\end{aligned}$$

For, Redwood Viscometer-II,

$$\begin{aligned}
A &= 0.002458 \\
B &= 0.4, \; T_r < 20s
\end{aligned}$$

Usually, at least three values are taken to take care of the discrepancies in readings.

As an alternative to kinematic viscosity, dynamic viscosity can also be measured using Brookfield Viscometers (Amrita, et al., 2014).

4.2.3 THERMAL CONDUCTIVITY

High cutting temperatures produced during machining damage the tool and the product quality. The major role of the cutting fluid in machining is to carry away the heat from the secondary shear zone in machining (Childs, 2000). All liquids, except water, have lower thermal conductivity at higher temperatures. Since water is a major ingredient of the water-soluble cutting fluids, it is important that the fluids have high thermal conductivity. To compare the efficacy of cutting fluids, the thermal conductivity of the cutting fluids is measured using thermal conductivity apparatus for liquids (Thomas & Sobhan, 2011). The test is usually done using either guarded hot-plate method (Johra, 2019; Xaman et al., 2009) or hot-wire method (Alvarado et al., 2012). Gaurded hot-plate apparatus consists of guarded hot plate assembly and control unit (ASTM, 2005a). After adding the sample into the apparatus, three heating coils and thermocouple belt are connected to the apparatus. The coils are turned on for heating the liquid. Once steady state was attained, readings from the thermocouples, voltmeter, and ammeter are recorded (Figure 4.6). Thermal conductivity of the sample is computed from the recorded steady-state temperatures for different power inputs using Fourier's conduction law. The apparatus basically operates on the principle that under heating from the top of the liquid, the heat transfer is only due to conduction. The calibration of the equipment may be done using two standard liquids like distilled water and glycerin (Figure 4.7–4.10).

The hot wire method is a measurement technique based on the measurement of the increase in temperature for a known height of liquid due to heating with a hot

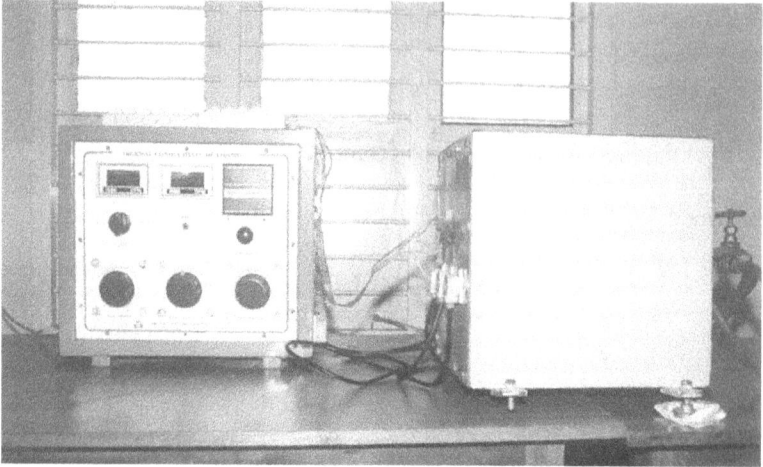

FIGURE 4.6   Guarded hot-plate apparatus (Revuru, 2008).

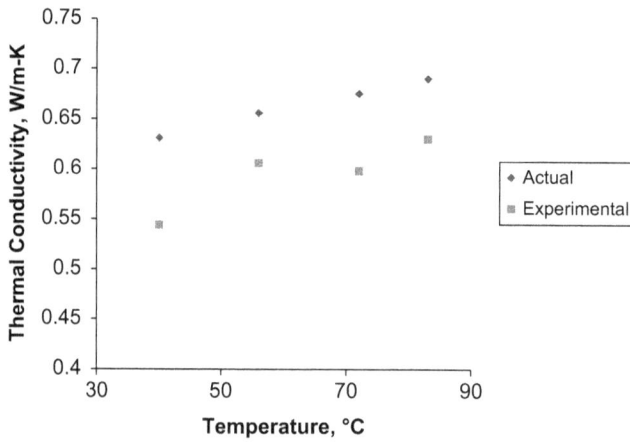

FIGURE 4.7   Calibration with thermal conductivity of water (Revuru, 2008).

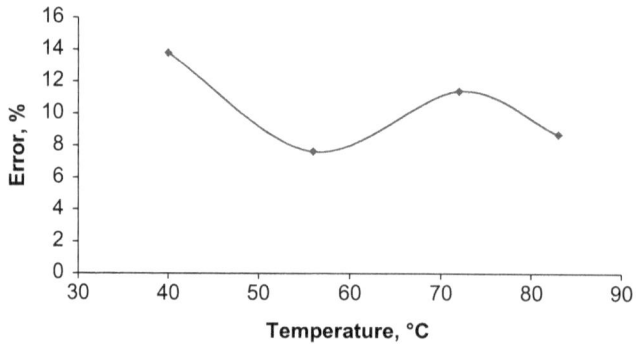

FIGURE 4.8   Error in measurement of thermal conductivity of water (Revuru, 2008).

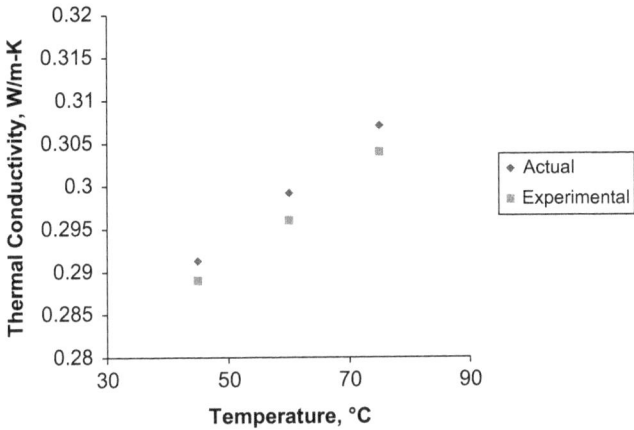

FIGURE 4.9   Validation with thermal conductivity of glycerin (Revuru, 2008).

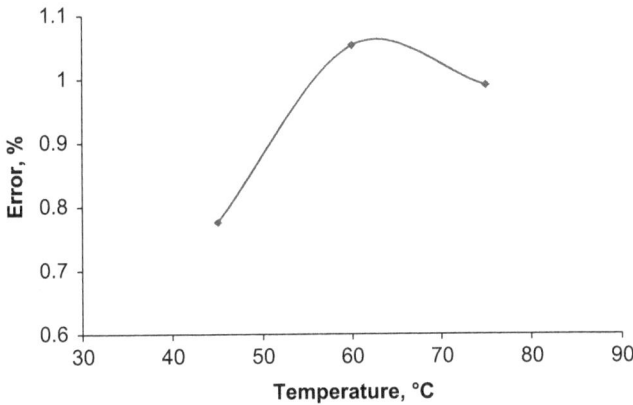

FIGURE 4.10   Error in measurement of thermal conductivity of glycerin (Revuru, 2008).

wire. The hot wire is dipped in the liquid and heated. The thermal conductivity of the liquid is calculated based on the heat supplied, time, and temperature rise. The thermocouple and the wire are insulated to prevent any contamination from the test fluid (Figure 4.11).

4.2.4   FLASH AND FIRE POINTS

Flash and fire points of the fluid are significant in high temperature applications. However, diluted fluids are not much prone to ignition due to the presence of water. As oil-emulsifier concentrates, without water, are sometimes used directly, determination of their flash and fire points is necessary. The tests are carried out in a Cleveland open cup tester (ASTM 2005b).

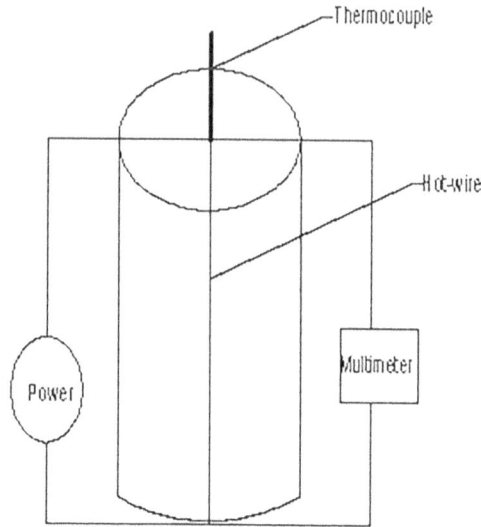

FIGURE 4.11   Schematic representation of transient hot-wire method.

4.2.5   pH VALUE

pH value helps to understand the condition of the fluid. A low pH value indicates the deterioration of the fluid. Also, a low pH value of the fluid is shown to increase tool wear. Too high pH values are hazardous to the operator and waste disposal. Also, pH value is used to understand the resistance of a fluid to microbial contamination. pH values for cutting fluids need to be maintained around 8.5–9.5 (Byers, 2017) to avoid corrosion of steels. pH value acts as a simple check to test the condition of the cutting fluid. pH values can be measured using digital pH indicator or pH strips.

4.2.6   RELATIVE DENSITY

Relative density helps to find out the concentration of a particular substance in a given solution. Water-miscible cutting fluids are formulated by mixing the emulsion with water. It is important to maintain the concentration of the fluid so that the flow and heating characteristics are not altered. Since the cutting fluid is continuously removing heat from the machining zone, some of the water is evaporated. This increases the concentration of the emulsion in the fluid. In order to maintain the performance of the fluid, it is important to measure the relative density and top-up with water. The relative density or specific gravity of the fluid can be found out using a hydrometer. It can be used to find out the dilution of the cutting fluid.

4.2.7   FOAMING

Since the cutting fluid is constantly circulated during machining, it is important that the fluid not exhibit excessive foaming due to the constant motion. Foaming not only impairs the visibility in the machining zone, but also affects the performance of the fluid as the flow is altered. Hence, it is important to control/eliminate foaming. The following test may be used to see the foaming characteristics (ASTM, 2018). In the first sequence, the small sample is taken in a foaming cylinder and the temperature is maintained at 24°C. Air is blown on the sample for some time and then after stopping the air, the amount of foam is measured. After a settling time of 10 minutes, the foam is measured for the second time. In the second sequence, the test is done identically but at 93.5°C. Then, the sample is cooled down to 43.5°C and placed in the cylinder. Again, the process is repeated.

4.2.8 CORROSION CHARACTERISTICS

Since the cutting fluids are commonly for machining steels and are constantly circulated in the machine tools and the cutting fluids contain water, the machine tools/workpiece may be subjected to corrosion. In fact, corrosion is one of the important disadvantages of water being used as a cutting fluid. Further, use of different compounds in the cutting fluid will lead to accelerated corrosion. Hence, it is important to evaluate the corrosion characteristics of the fluids. Also, if the fluid is non-corrosive, then the use of corrosion-inhibitors can be eliminated. The test for corrosion is done using potentiodynamic polarization (ASTM, 2014). Higher corrosion resistance is indicated by higher value of the potential. The test basically assesses the corrosion of a metal in an environment. In this test, a small potential scan is given to the sample that is exposed to the cutting fluids and current versus potential curve is plotted. Steeper curve indicates better corrosion resistance (Figure 4.12).

Another test that is commonly used is cast iron chip test (Figure 4.13) (ASTM D4627; ASTM D4627-12, 2017). Cast iron chips are taken on filter paper

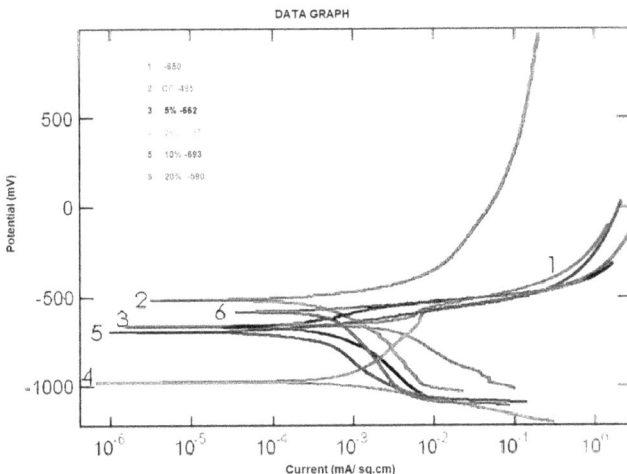

FIGURE 4.12 Corrosion test results (Srikant & Ramana, 2016).

and soaked in 10 mL of the cutting fluid for four hours. At the end of the test duration, the chips are removed from the filter paper. If the filter paper contains any rust spots, then the fluid is corrosive. This test can be used to find out the minimum concentration of the fluid to avoid corrosion (Astakhov and Joksch, 2012).

4.3 MICROBIAL CONTAMINATION

Due to the stringent environmental regulations, maintenance and disposal of cutting fluids are critical. Cutting fluids contain complex hydrocarbon compounds like oils and emulsifiers dispersed in water. These compounds are highly vulnerable to microbial contamination (Shokoohi et al., 2015). The organisms that are developed in the cutting fluids are classified as aerobic bacteria, anaerobic bacteria, and fungi (Rossmoore, 1981).

Microbial contamination of the fluids plays a key role in the deterioration of the fluids over time, leading to poor performance of the fluid, health issues for the operators, and poor working conditions in the shop floor. Hence, resistance to contamination is a very important criterion for the selection of the fluids. The disposal of contaminated cutting fluid can cause water and soil pollution (Sokovic and Mijanović, 2001). Treatment of the used fluids is an additional cost to the industry (Davim, 2013). During usage, the bacteria can enter into the bodies of the shop floor workers through cuts and other open injuries (Figure 4.14).

The bacteria can spread fast and lead to amputation of limbs. Hence, the assessment of contamination and identification of the bacteria is important (Suliman et al., 1997).

Greeley et al. (2004) developed a drill test bed that could detect small changes in metal working fluid composition by monitoring the changes in thrust, torque, and drill temperature. It was reported that the degradation in the function of the

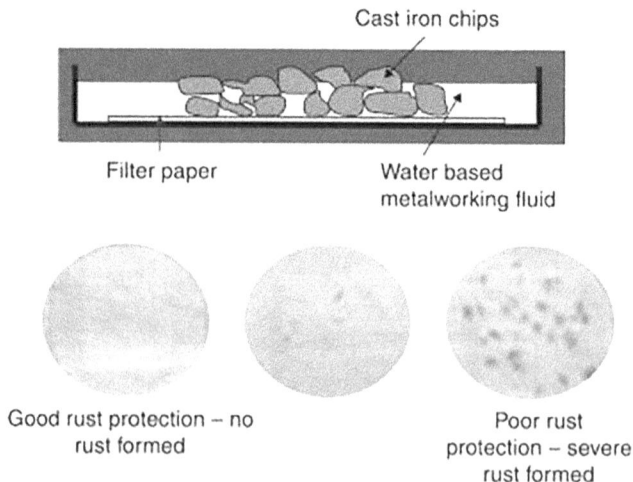

Cast iron chips

Filter paper   Water based
                      metalworking fluid

Good rust protection – no   Poor rust
rust formed   protection – severe
                                   rust formed

FIGURE 4.13   Cast iron chip test for corrosion testing of fluids (Evans, 2012).

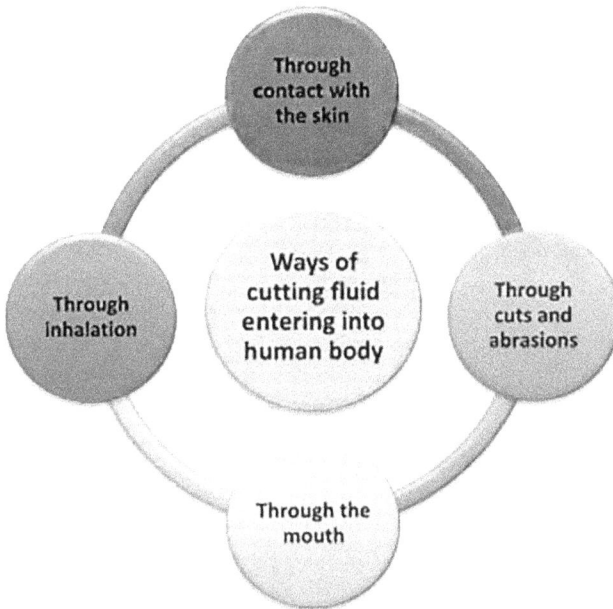

FIGURE 4.14 The common ways of metal-working fluid entering into the human body (Haider & Hashmi, 2014).

lubricant is due to depletion in synthetic and semi-synthetic metal working fluid. However, significant changes were noted only for larger percentages of component depletion. The torque response did not indicate any change in the functionality of synthetic metalworking fluids.

The growth of aerobic bacteria leads to separation of the emulsion, corrosion, and loss of lubricity as this type of bacteria is extremely oxidative and is adoptable to the wide variety of organic molecules which are found in the metal cutting fluids (Figure 4.15). Further, the operators are constantly exposed to the cutting fluids that cause skin diseases, resulting in folliculitis. Because of the use of these cutting fluids, they also become vulnerable to skin disease like infections, oil acne, allergic dermatitis, and mechanical trauma (Rossmoore, 1981) and the throat, lungs, and nose are severely affected by these diseases (Iowa Waste Reduction Center, 1996).

The plate count method is commonly used in estimating the microbial contamination (Cyprowski, 2012; Saha & Donofrio, 2012). Petri plates with nutrients (usually agar) are used in this test. These nutrients provide the required medium for the growth of bacterial colonies. The choice of nutrient in the petri plates is based on the nature of the bacteria: aerobic/anerobic (Rossmoore, 1995). A dye may also be added to help in counting the colonies. The test fluid is usually diluted with water in the ratio of 1:100. 0.1 mL of the prepared sample is collected on a petri plate under the sterile laminar air flow. There is a blue flame in the cabinet that prevents contamination of the sample while it is being transferred to the petri plate. The sample is then incubated in a B.O.D incubator at 37°C for 24–48 hours. The

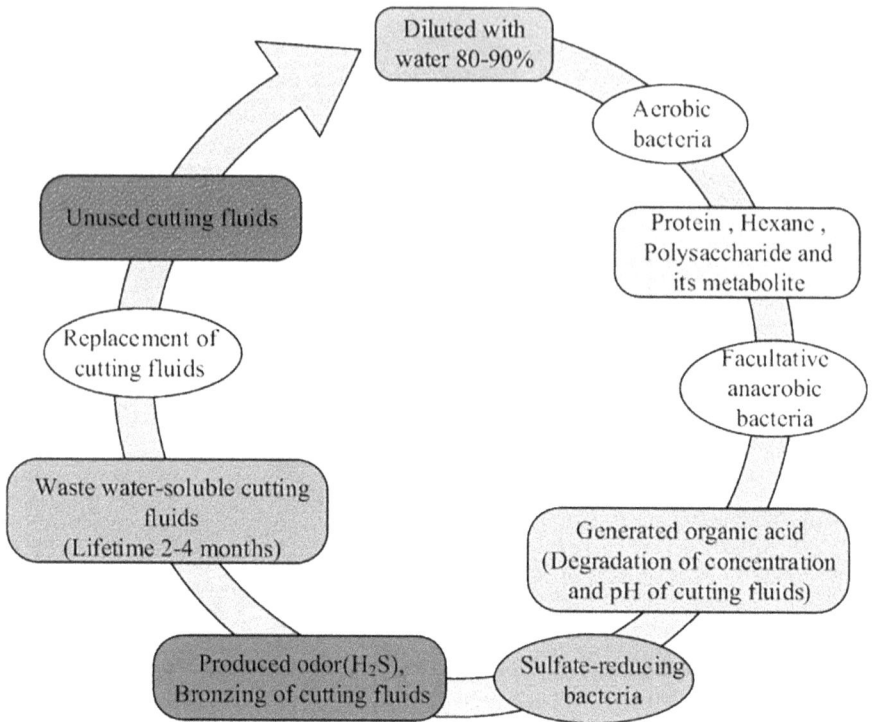

FIGURE 4.15   Microbial degradation of cutting fluids (Ma et al., 2018).

bacterial colonies are counted after incubation. The test is usually carried out in triplicate to rule out any anomalies in the process.

In order to identify the microbial species, the same process is repeated using petri plates with different nutrients that are specific to a particular type of microorganism. Other common tests like the oxidase test are also carried out (Harley, 2007). The species of the bacteria is identified based on the medium that produced colony growth (Table 4.1).

In general, the microbial species found in the cutting fluids belong to *Pseudomonas* genus (Rossmoore, 1981). The species exhibit the broadest appetite and minimum nutritional requirement than any other type of microorganisms and are difficult to kill. These bacteria are sometimes found even in the biocides are used in hospitals. *Pseudomonas* grows best under conditions of maximum aeration as they are highly oxidative and their growth rate is typically multiplying every 45 minutes under conditions of the cutting fluid. This type of microorganism is highly opportunistic and non-invasive, but whoever has cuts or wounds is common in any machine shop and can easily be infected.

Due to the regulations made by various government agencies, like the Environmental Protection Agency (EPA), disposal of the used oil is expensive. Hence, the use of biocides is prohibitive, though they are the most promising solutions to the contamination of the cutting fluids.

TABLE 4.1

Frequency and percentage of occurrence of bacteria and fungi in cutting fluids (Suliman et al., 1997)

Isolate	Before circul.	During circul.	After circul.	Emulsion	Number	Percentage
Aspergillus species	6	5	7	1	19	24·36
Pseudomonas aeruginosa	5	3	4	1	13	16·67
Citrobacter diversus	4	4	4	—	12	15·38
Pseudo. fluorescens putida	4	1	4	2	11	14·10
Enterobacter sakazakii	2	1	2	—	5	6·41
Pseudomonas maltophilia	1	—	2	2	5	6·41
Pseudomonas species	1	—	1	3	5	6·41
Microsporum species	1	—	1	1	3	3·85
Staphylococcus xylosus 2	—	—	2	—	2	2·56
Micrococcus varians rosous	—	—	1	—	1	1·28
Staphylococcus capitis	—	—	—	1	1	1·28
Citrobacter frundii	1	—	—	—	1	1·28
Total/percentage	25	14	28	11	78	100

4.4 CONCLUSION

Since cutting fluids can be of various compositions and additives, it is not possible to generalize the behavior. In order to understand the behavior and improve the performance of the cutting fluids, it is important to assess the basic properties of the cutting fluids. Various properties need to be assessed depending on the priorities set of the fluid based on the application. Standard tests need to be followed to estimate the properties of the cutting fluids, as explained in this chapter.

REFERENCES

Alvarado, S., Marín, E., Juárez, A. G., Calderón, A., & Ivanov, R. (2012). A hot-wire method based thermal conductivity measurement apparatus for teaching purposes. *European Journal of Physics*, 33, 897–906. 10.1088/0143-0807/33/4/897.

Amrita, M., Srikant, R., Sitaramaraju, A., Prasad, M., & Krishna, P. V. (2014). Preparation and characterization of properties of nanographite-based cutting fluid for machining operations. *Proceedings of the Institution of Mechanical Engineers, Part J: Journal of Engineering Tribology*, 228(3), 243–252. 10.1177/1350650113502003

Astakhov, V. P., & Joksch, S. (Eds.). (2012). *Metalworking fluids (MWFs) for cutting and grinding: fundamentals and recent advances.* UK: Woodhead Publishing.

ASTM D4627-12 (2017). Standard Test Method for Iron Chip Corrosion for Water–Miscible Metalworking Fluids, 05.02.

ASTM G59-97 (2014) *Standard test method for conducting potentiodynamic polarization resistance measurements*. West Conshohocken, PA: ASTM International.

ASTM Standard C177–04. (2005a). *Test method for steady-state heat flux measurements and thermal transmission properties by means of the guarded-hot-plate apparatus*, ASTM Book of Standards, 04.06, Construction, thermal insulation, environmental acoustics. Philadelphia, PA: ASTM International.

ASTM Standard D0092–02B. (2005b). *Test method for flash and fire points by Cleveland open cup tester*, ASTM Book of Standards, 05.01, Petroleum products and lubricants. Philadelphia, PA: ASTM International.

ASTM Standard D0445–04E01. (2005). *Test method for kinematic viscosity of transparent and opaque liquids*, ASTM Book of Standards, 05.01, Petroleum products and lubricants. Philadelphia, PA: ASTM International.

ASTM D1401-21. (2021). "Standard Test Method for Water Separability of Petroleum Oils and Synthetic Fluids", 05.01.

ASTM, D. 892 (2018). *Standard test method for foaming characteristics of lubricating oils*. American Society for Testing and Materials. International.

Byers J. P. (2017). *Metalworking fluids*, 3rd ed.

Cutting Fluid Management in Small Machining Operations, IOWA waste reduction Center, University of Northern IOWA, IOWA, (1996).

Childs, T. H. (2000). *Metal machining: theory and applications*. USA: Butterworth-Heinemann.

Cyprowski, M. (2012). Microbial contamination of metalworking fluids. *Bezp Pr Nauk Prakt*, 9, 16–19.

Davim, J. P. (Ed.). (2013). *Green manufacturing processes and systems* (pp. 1–22). Berlin: Springer.

Evans, R. (2012). Selection and testing of metalworking fluids. In *Metalworking Fluids (MWFs) for Cutting and Grinding* (pp. 23–78). Woodhead Publishing.

Greeley, M. H., Devor, R. E., Kappoor, S. G., & Rajagopalan, N. (2004). The influence of fluid management policy and operational changes on metal working fluid functionality. *Journal of Manufacturing Science and Engineering*, 126, 445–450.

Haider, J., & Hashmi, M. S. J. (2014). 8.02—Health and environmental impacts in metal machining processes. *Comprehensive Materials Processing*, 8, 7–33.

Harley, J. P. (2007). *Microbiology lab manual*. McGraw-Hill Asia.

Johra, H. (2019). *Description of the Guarded Hot Plate Method for thermal conductivity measurement with the EP500*, Denmark: Aalborg University. 10.54337/aau317020205

Lawal, S. A., Ugheoke, B. I., Woma, T. Y., Ikporo, J. U., Ogundare, T. A., Nonye, C., & Okoye, I. G. (2015). Effect of emulsifier content on the properties of vegetable oil based cutting fluid. *American Journal of Materials Engineering and Technology*, 3 (3), 63–69. doi: 10.12691/materials-3-3-3.

Ma, S., Kim, K., Huh, J., Kim, D. E., Lee, S., & Hong, Y. (2018). Regeneration and purification of water-soluble cutting fluid through ozone treatment using an air dielectric barrier discharge. *Separation and Purification Technology*, 199, 289–297

Revuru, R. S. (2008). *Experimental investigations on the influence of emulsifier content in cutting fluids on machining performance and occupational environment*, Doctoral thesis. Andhra University.

Rossmoore, H. W. (1981). Antimicrobial agents for water-based metalworking fluids. *Journal of Occupational and Environmental Medicine*, 23(4), 247–254.

Rossmoore, H. W. (1995). Microbiology of metalworking fluids: Deterioration, disease and disposal. *J Soc Triobio Lub Eng*, 51, 112–118

Saha, R., & Donofrio, R. S. (2012). The microbiology of metalworking fluids. *Appl Microbiol Biotechnol*, 94, 1119–1130. 10.1007/s00253-012-4055-7

Shokoohi, Y., Khosrojerdi, E., & Shiadhi, B. R. (2015). Machining and ecological effects of a new developed cutting fluid in combination with different cooling techniques on turning operation. *Journal of Cleaner Production*, 94, 330–339.

Soković, M., & Mijanović, K. (2001). Ecological aspects of the cutting fluids and its influence on quantifiable parameters of the cutting processes. *Journal of Materials Processing Technology*, 109(1–2), 181–189.

Srikant, R., & Ramana, V. V. (2016). Studies on corrosion and quenching effects of cutting fluid with vegetable-based emulsifier on AISI 1040 steel. *Proceedings of the Institution of Mechanical Engineers, Part B: Journal of Engineering Manufacture*, 230(1), 178–181. 10.1177/0954405415597621

Suliman, S. M. A., Abubakr, M. I., & Mirghani, E. F. (1997). Microbial contamination of cutting fluids and associated hazards. *Tribology International*, 30(10), 753–757.

Thomas, & Sobhan, B. P. (2011). Nanoscale Research Letters, 6, 377. http://www.nanoscaleslett.com/content/6/1/377

Xamán, J., Lira, L., & Arce, J. (2009). Analysis of the temperature distribution in a guarded hot plate apparatus for measuring thermal conductivity. *Applied Thermal Engineering*, 29(4), 617–623. 10.1016/j.applthermaleng.2008.03.033

5 MQL equipment and characterization of spray

5.1 INTRODUCTION

Though cutting fluids are being used in machining for cooling and lubrication, the use of cutting fluids is associated with various environmental problems and costs. In order to eliminate the drawbacks of the cutting fluids, the use of cutting fluids needs to be minimized in terms of the quantities of the fluid used. Many strategies are developed to limit the amount of cutting fluid used in machining. One of the popular techniques is minimum quantity lubrication (MQL).

5.2 MINIMUM QUANTITY LUBRICATION

MQL replaces the traditional flood lubrication by applying mist of the fluid in place of large quantities of the cutting fluid. The amount of fluid applied drastically decreases from about 100 L/hr to 50–60 mL/hr. This has gained importance in recent years due to the promising results that were observed and increasing awareness on sustainable machining. Another advantage of MQL is the reduction in cost due to the very small amounts of fluid used and no disposal. Though mostly straight oils are used in MQL, some researchers have studied the application of water-miscible oils. While the neat oils provide better lubrication and corrosion protection, they are poor in cooling and are flammable. The water-miscible oils provide good lubrication, cooling, non-flammable, but have the problem of microbial contamination. Similarly, semi-synthetic and synthetic fluids are good with cooling but can be easily polluted (Singh et al., 2020). In order to improve the performance of the fluid in MQL, the addition of nanoparticles is gaining popularity (Hamran et al., 2020). Many studies can be found in the literature that aim to identify the optimal MQL fluid for machining different materials. Details of several studies on MQL can be found in Chapter-8.

The reason for using mist/aerosols is increasing the surface contact area of the fluid with the machining zone to help in better lubrication and cooling. It must be noted that while the flood lubrication mainly concentrates on the removal of heat by conduction, MQL concentrates on lubricating the tool/chip interface to reduce the generation of heat. MQL is not very effective in carrying away the heat compared to the flood lubrication, due to the lower thermal conductivity of the neat oils than the water-miscible oils.

Since MQL's main aim is to reduce the friction, the method and direction of application are critical. Cutting fluids, in general, are applied in one or more of the following directions (Figure 5.1):

 DOI: 10.1201/9781003328742-5

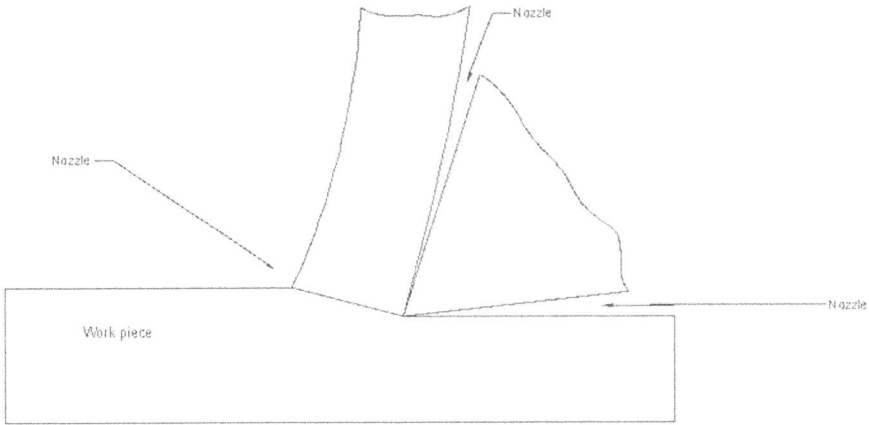

FIGURE 5.1 Directions of application of cutting fluids (redrawn from Revuru et al., 2017).

   a. on the rake face of the tool
   b. tool/chip interface
   c. on the underside of the chip

Tasdelen et al. (2008) applied the fluid in MQL, compressed air, and water emulsion on the rake face of the tool. It was found that the emulsion had better performance as indicated by the lowest tool/chip contact length. This is because when the coolant is applied on the rake face, the main function becomes conduction of heat rather than lubrication. In many cases, it is not possible for the cutting fluid to reach the tool/chip interface.

Banerjee and Sharma (2019) studied the application of MQL in two different directions: on the rake face and directed to the tool/chip interface. It was found that the application of MQL at multiple points is more beneficial compared to a single-point application.

Childs et al. (2000) pointed out that the application of the cutting fluid at the underside of the chip is most beneficial as the fluid can reach the tool/chip interface by capillary through the micro-holes formed in the chips.

In order to increase the chances of the fluid to reach the tool/chip contact area and spread as a lubricating film, it is necessary to ensure that the size of the particles is small and the contact area with the tool is large. Special equipment is used to accomplish these requirements.

5.3   EQUIPMENT FOR MQL

It is important to supply cutting fluids as aerosol particles in MQL. This is to ensure that the surface contact is more and the full potential of the fluid is utilized to lubricate the machining zone. In order to achieve this, special pieces of equipment are required.

MQL systems can be categorized into two categories: external application and internal application (Boswell et al., 2017). Figure 5.2 shows the two methods of

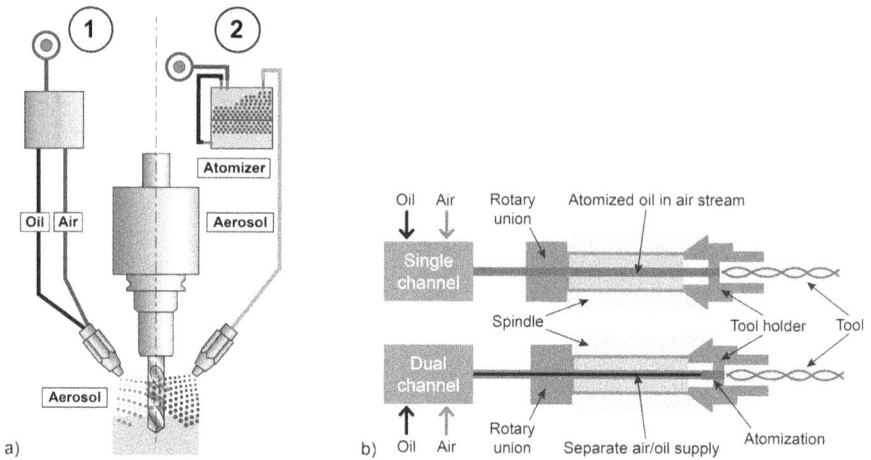

FIGURE 5.2   Schematic representation of a) external application using ejector nozzle and conventional nozzle, b) internal application (Boswell et al., 2017).

application. Internal application is popularly used in applications like drilling, while external application is used in milling, turning, grinding, etc. In the ejector nozzle, oil and compressed air are supplied separately and mixed before supplying the fluid to the machining zone. Whereas in the conventional nozzle, the compressed air and oil are mixed in an atomizer and the aerosol spray is supplied to the nozzle. In the single channel for internal MQL, oil and compressed air are mixed before supplying, while in the dual channel, the oil and compressed air are supplied separately and mixed just before supplying them to the tool.

As the jet of aerosol particles leaves the nozzle, it takes the form of a conical spray. The inner part of the spray contains the high mass oil and larger drops while the outer surface has smaller drops (Hadad & Beigi, 2021). Hence, the velocity of the jet is highest at the center and decreases at the outer part of the cone. It was found that the MQL spray angle of 10–12° is most effective (Figure 5.3).

As more fluids are introduced into the MQL system, the complexity of the nozzle changes. If more fluids like liquid CO_2 (cryogenic) are added, then a Coanda effect nozzle is used (Figure 5.4).

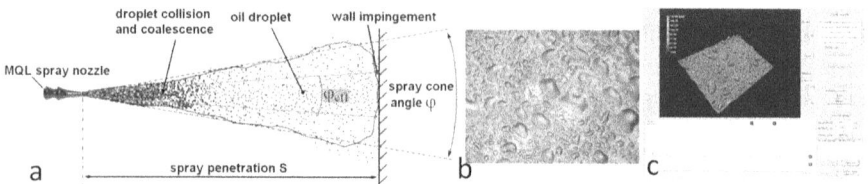

FIGURE 5.3   a) Schematic representation of MQL spray angle = 10–12°, b) MQL oil droplets on the target surface, c) 3D image analysis of droplets on the surface (Hadad & Hadi, 2013).

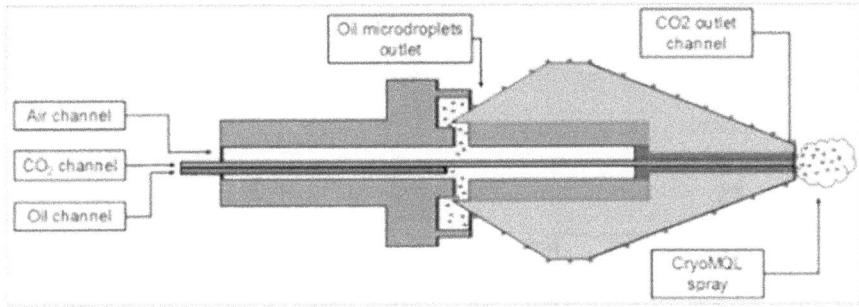

FIGURE 5.4   Coanda effect nozzle (Pereira et al., 2017).

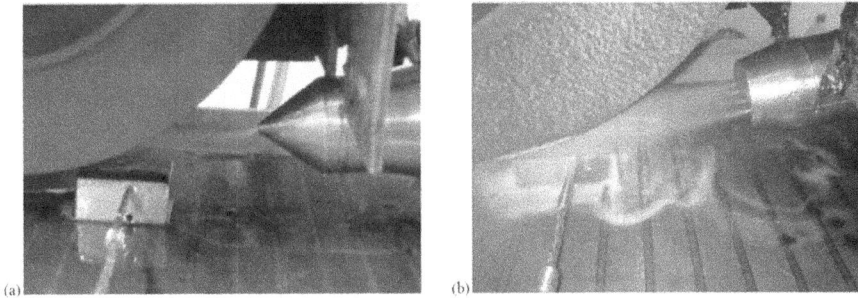

FIGURE 5.5   a) Conventional nozzle, b) modified nozzle (Sharmin et al., 2021).

In this nozzle, the oil is made into an aerosol spray and then it combines with the cryogenic gas before reaching the machining zone. Pereira et al. (2017) used computational fluid dynamics to optimize the size of the nozzle. In an interesting study, Sharmin et al. (2021) designed a nozzle with multiple holes. A 2-mm diameter conventional nozzle was modified to have four holes of 0.5-mm diameter (Figure 5.5). A considerable reduction in cutting temperatures was observed.

Revuru et al. (2018) applied MQL in the turning of Ti6Al4V. Soybean oil was supplied at the underside of the chip. Cutting forces, surface roughness, and tool wear were measured at regular intervals (Figure 5.6).

As mentioned earlier, nanofluids were used by some researchers in MQL. In order to prevent agglomeration of the nanoparticles in the fluid, ultrasonic-assisted oscillators were used (Figure 5.7).

5.4   CHARACTERIZATION OF MQL

Since very small quantities of the fluid are supplied in MQL, the selection of cutting fluids and proper application is very important. In order for the fluids to provide the required cooling and lubrication, it is necessary to ensure that the fluids reach the tool/chip interface and adhere to the surface. Apart from the characterization of the properties of the fluids, as described in earlier chapters, it is also important to characterize some additional properties of the MQL fluids.

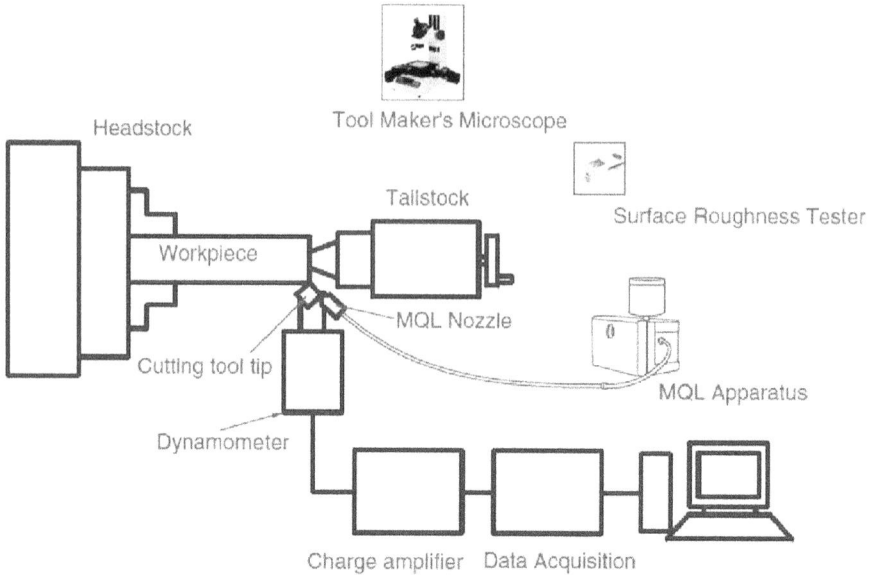

FIGURE 5.6   Application of cutting fluids in turning (Revuru et al., 2018).

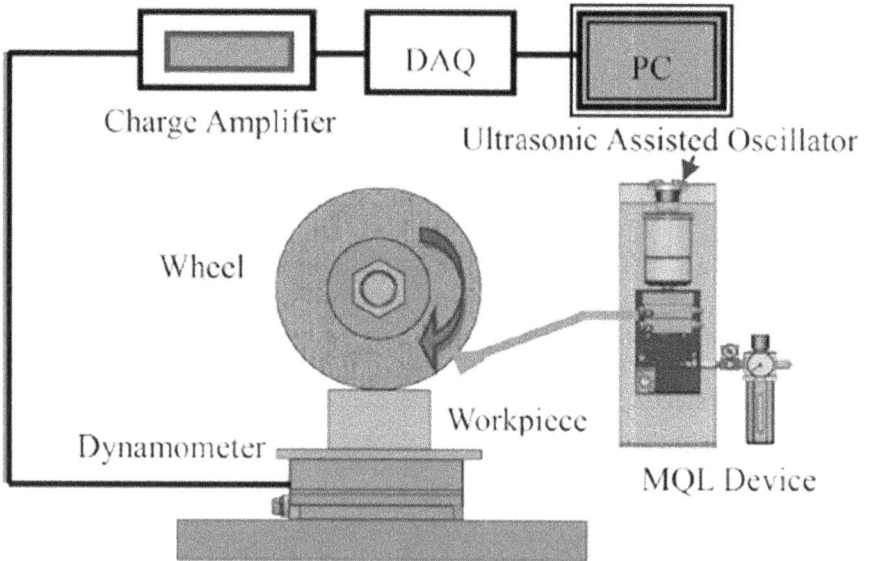

FIGURE 5.7   Use of ultrasonic-assisted oscillators (Huang & Liu, 2016).

5.4.1   SURFACE TENSION

Surface tension is the tendency of liquid to assume minimum surface area. This, in turn, reduces the ability of the fluid to spread on the surface, thus reducing the lubricating abilities. Surface tension affects the wettability of a liquid. Surface tension is commonly measured using DuNouy Ring and Wilhelmy plate methods (Ebnesajjad, 2011). In both of these techniques, a droplet is placed on a surface and pulled with a known geometry: a ring/plate. The pulling force is measured and surface tension is calculated. Other commercially available equipment include force tensiometers (Kaluarachchi et al., 2021; ASTM D971-20).

5.4.2 WETTABILITY

Wettability of a liquid is critical to understand the capillarity action of the fluid. In order to find the wettability, the contact angle between the liquid and the surface needs to be measured (Figure 5.8). The higher the contact angle, the lower the wetting of the surface. The Amott method is commonly used for the measurement of contact angle (McPhee et al., 2015).

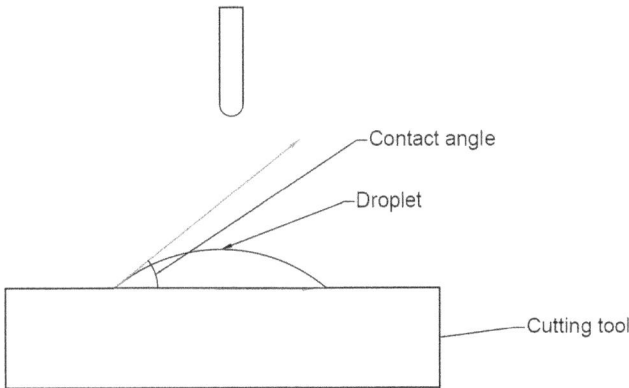

FIGURE 5.8   Contact angle of a droplet (Su et al., 2016).

This method uses special but inexpensive equipment, as shown in Figure 5.9. The volume of non-imbibded oil/brine solution is measured when the sample is placed in the equipment. This is spontaneous imbibing. Also, the sample can have ford imbibing by centrifuging. Another important technique known as the USBM test (U.S. Bureau of Mines) also uses a similar technique of centrifuging (Abdallah et al., 2007).

Another technique is the sessile drop method (McPhee et al., 2015). In this method, a drop of the fluid is taken on the sample plate and then studied under the contact angle goniometer to measure the contact angle (Figure 5.10).

5.4.3 AEROSOL DROPLET SIZE

Aerosol droplet size is dependent on the surface tension, wettability of the liquid, and pressure of the compressed air supplied in the MQL equipment. A smaller droplet size is beneficial for the capillary action and uniform wetting of the surface.

Aerosol droplet size can be measured using the laser diffraction technique (Fritz & Hoffmann, 2016). As the droplets pass through the laser beam, the information

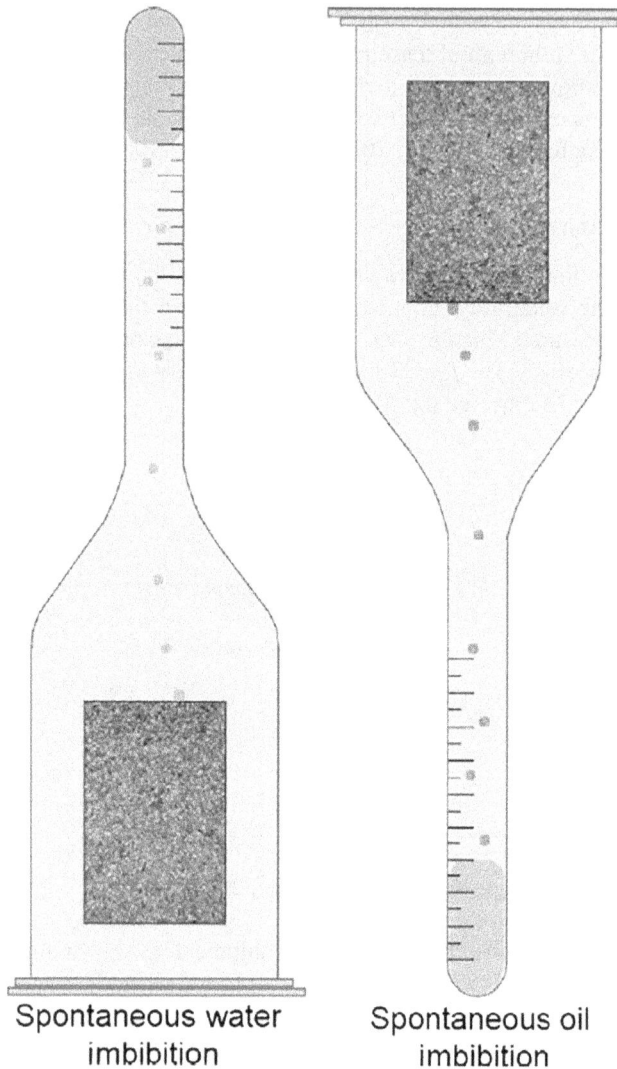

FIGURE 5.9 Wettability measurement (McPhee et al., 2015).

regarding the size of the droplets is collected (Figure 5.11). One problem with this method is that sometimes cluster sizes may be measured instead of the actual droplet size. McMurry (2000) provides a list of different instruments available for measuring the droplet size (Figure 5.12).

Another method that is commercially used is the time domain-nuclear magnetic resonance (TD-NMR) method. In this technique, the movement of the droplets is restricted, instead of having a bulk flow. This makes it possible to collect information on each droplet. (Oxford Instruments, 2023).

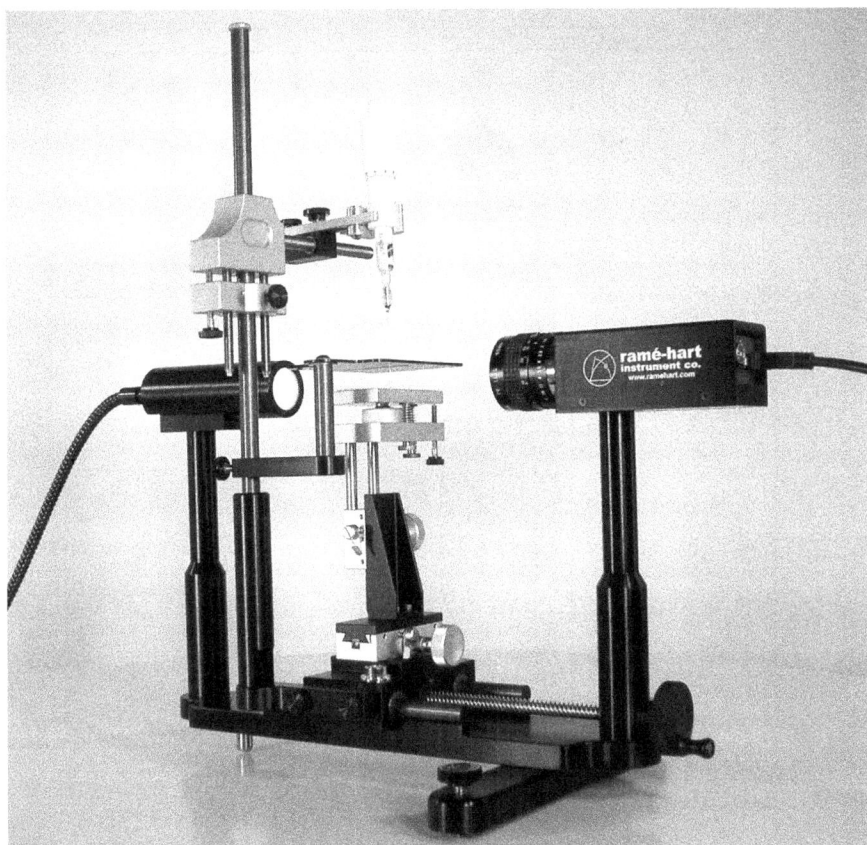

FIGURE 5.10 Contact angle goniometer ("File:Rame-hart goniometer.jpg" by Ramehart is licensed under CC BY 3.0.).

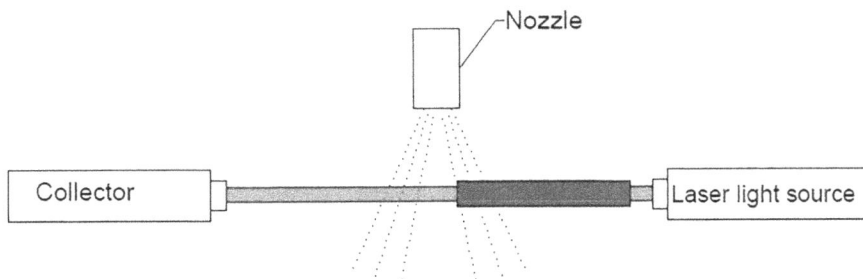

FIGURE 5.11 Laser-diffraction principle: Malvern Spraytec.

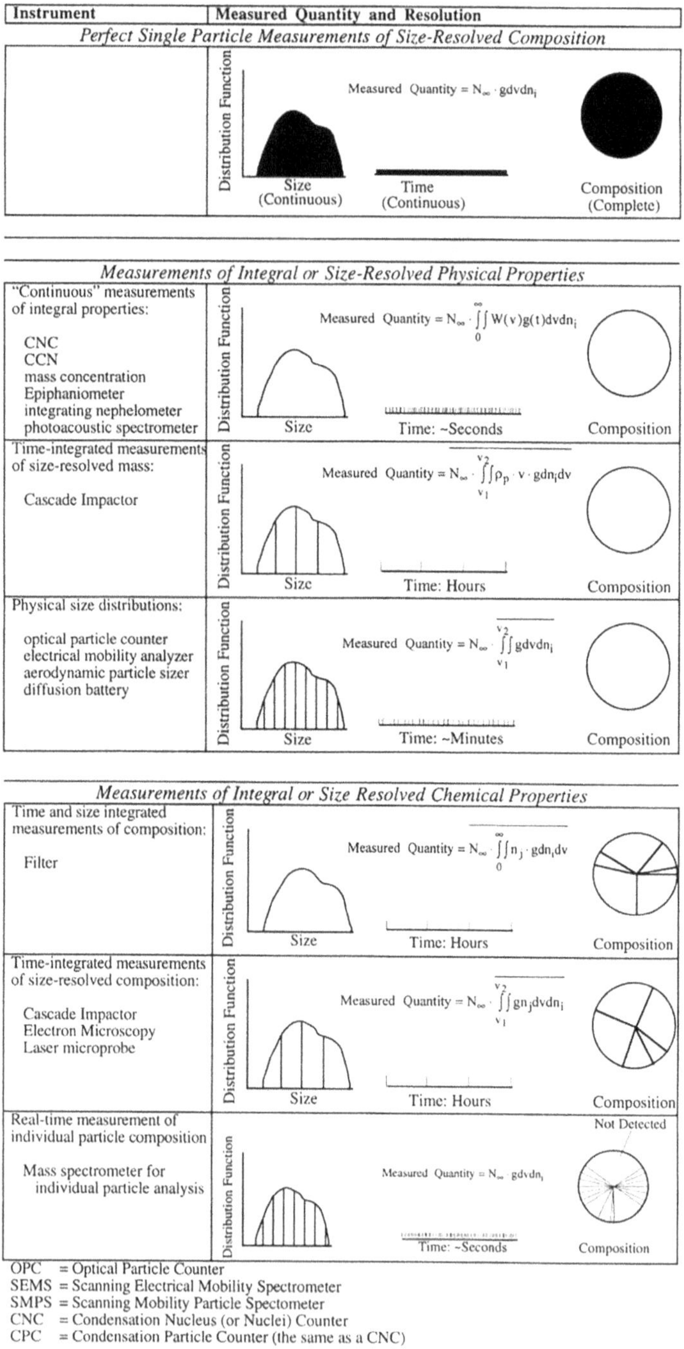

Instrument	Measured Quantity and Resolution

Perfect Single Particle Measurements of Size-Resolved Composition

Measured Quantity $= N_\infty \cdot g\,dv\,dn_i$

Size (Continuous) — Time (Continuous) — Composition (Complete)

Measurements of Integral or Size-Resolved Physical Properties

"Continuous" measurements of integral properties:

CNC
CCN
mass concentration
Epiphaniometer
integrating nephelometer
photoacoustic spectrometer

Measured Quantity $= N_\infty \cdot \int\int_0^\infty W(v)g(t)dvdn_i$

Size — Time: ~Seconds — Composition

Time-integrated measurements of size-resolved mass:

Cascade Impactor

Measured Quantity $= N_\infty \cdot \int\int_{v_1}^{v_2}\rho_p \cdot v \cdot gdn_i dv$

Size — Time: Hours — Composition

Physical size distributions:

optical particle counter
electrical mobility analyzer
aerodynamic particle sizer
diffusion battery

Measured Quantity $= N_\infty \cdot \int\int_{v_1}^{v_2} gdvdn_i$

Size — Time: ~Minutes — Composition

Measurements of Integral or Size Resolved Chemical Properties

Time and size integrated measurements of composition:

Filter

Measured Quantity $= N_\infty \int\int_0^\infty n_j \cdot gdn_j dv$

Size — Time: Hours — Composition

Time-integrated measurements of size-resolved composition:

Cascade Impactor
Electron Microscopy
Laser microprobe

Measured Quantity $= N_\infty \int\int_{v_1}^{v_2} gn_j dvdn_i$

Size — Time: Hours — Composition

Real-time measurement of individual particle composition

Mass spectrometer for individual particle analysis

Not Detected

Measured Quantity $= N_\infty \; gdvdn_i$

Time: ~Seconds — Composition

OPC   = Optical Particle Counter
SEMS = Scanning Electrical Mobility Spectrometer
SMPS = Scanning Mobility Particle Spectometer
CNC   = Condensation Nucleus (or Nuclei) Counter
CPC   = Condensation Particle Counter (the same as a CNC)

FIGURE 5.12   Classification of aerosol instruments (McMurry, 2000).

5.5 CONCLUSION

MQL delivers aerosol droplets to the machining zone. Since very little fluid is delivered, it is important to maximize the efficacy of the fluid. Apart from the properties of the fluid, the characterization of the aerosol particles is also important. This chapter provides details about the different techniques used to deliver and characterize the different aspects of MQL lubricant.

REFERENCES

Abdallah, W., Buckley, J. S., Carnegie, A., Edwards, J., Herod, B., Fordham, E., Graue, A., Habashy, T., Seleznev, N., Signer, C., Hussain, H., Montaron, B., & Ziauddin, M. (2007). Fundamentals of wettability. *Oil Field Review*, 44–61.

ASTM D971-20 Standard Test Method for Interfacial Tension of Insulating Liquids Against Water by the Ring Method, 10.03.

Banerjee, N., & Sharma, A. (2019). Improving machining performance of Ti-6Al-4V through multi-point minimum quantity lubrication method. *Proceedings of the Institution of Mechanical Engineers, Part B: Journal of Engineering Manufacture*, 233(1), 321–336.

Boswell, B., Islam, M. N., Davies, I. J., Ginting, Y. R., & Ong, A. K. (2017). A review identifying the effectiveness of minimum quantity lubrication (MQL) during conventional machining. *The International Journal of Advanced Manufacturing Technology*, 92, 321–340.

Childs, T. H., Maekawa, K., & Obikawa, T. (2000). *Metal machining: theory and applications*. Butterworth-Heinemann.

Ebnesajjad, S. (2011). Surface tension and its measurement. In *Handbook of adhesives and surface preparation* (pp. 21–30). William Andrew Publishing.

Fritz, B. K., & Hoffmann, W. C. (2016). Measuring spray droplet size from agricultural nozzles using laser diffraction. *Journal of Visualized Experiments: JoVE*, (115).

Hadad, M., & Beigi, M. (2021). A novel approach to improve environmentally friendly machining processes using ultrasonic nozzle–minimum quantity lubrication system. *The International Journal of Advanced Manufacturing Technology*, 114, 741–756.

Hadad, M., & Hadi, M. (2013). An investigation on surface grinding of hardened stainless steel S34700 and aluminum alloy AA6061 using minimum quantity of lubrication (MQL) technique. *The International Journal of Advanced Manufacturing Technology*, 68(9–12), 2145–2158.

Hamran, N. N., Ghani, J. A., Ramli, R., & Haron, C. C. (2020). A review on recent development of minimum quantity lubrication for sustainable machining. *Journal of Cleaner Production*, 268, 122165.

Huang, W. T., & Liu, W. S. (2016). Investigations into lubrication in grinding processes using MWCNTs nanofluids with ultrasonic-assisted dispersion. *Journal of Cleaner Production*, 137, 1553–1559.

Kaluarachchi, C. P., Lee, H. D., Lan, Y., Lansakara, T. I., & Tivanski, A. V. (2021). Surface Tension Measurements of Aqueous Liquid–Air Interfaces Probed with Microscopic Indentation. *Langmuir*, 37(7), 2457–2465.

McPhee, C., Reed, J., & Zubizarreta, I. (2015). Wettability and wettability tests. In *Developments in Petroleum Science* (Vol. 64, pp. 313–345). Elsevier.

McMurry, P. H. (2000). A review of atmospheric aerosol measurements. *Atmospheric Environment*, 34(12–14), 1959–1999.

Oxford Instruments. Measurement of droplet size distribution in emulsions using benchtop NMR. Available on https://nmr.oxinst.com/learning/view/article/measurement-of-droplet-size-distribution-in-emulsions-using-benchtop-nmr accessed on April, 04, 2023.

Pereira, O., Rodríguez, A., Barreiro, J., Fernández-Abia, A. I., & De Lacalle, L. N. L. (2017). Nozzle design for combined use of MQL and cryogenic gas in machining. *International Journal of Precision Engineering and Manufacturing-green Technology*, 4, 87–95.

Revuru, R. S., Posinasetti, N. R., & Vsn, V. R. (2017). Application of cutting fluids in machining of titanium alloys—a review. *The International Journal of Advanced Manufacturing Technology*, 91, 2477–2498.

Revuru, R. S., Zhang, J. Z., Posinasetti, N. R., & Kidd, T. (2018). Optimization of titanium alloys turning operation in varied cutting fluid conditions with multiple machining performance characteristics. *The International Journal of Advanced Manufacturing Technology*, 95, 1451–1463.

Sharmin, I., Moon, M., Talukder, S., Alam, M., & Ahmed, M. F. (2021). Impact of nozzle design on grinding temperature of hardened steel under MQL condition. *Materials Today: Proceedings*, 38, 3232–3237.

Singh, G., Gupta, M. K., Hegab, H., Khan, A. M., Song, Q., Liu, Z., … & Pruncu, C. I. (2020). Progress for sustainability in the mist assisted cooling techniques: a critical review. *The International Journal of Advanced Manufacturing Technology*, 109, 345–376.

Su, Y., Gong, L., Li, B., Liu, Z., & Chen, D. (2016). Performance evaluation of nanofluid MQL with vegetable-based oil and ester oil as base fluids in turning. *The International Journal of Advanced Manufacturing Technology*, 83, 2083–2089.

Tasdelen, B., Thordenberg, H., & Olofsson, D. (2008). An experimental investigation on contact length during minimum quantity lubrication (MQL) machining. *Journal of Materials Processing Technology*, 203(1–3), 221–231.

6 Minimum quantity lubrication (MQL) for different machining operations

6.1 INTRODUCTION

Nowadays, due to the strict environmental protection laws and huge cutting fluid costs, most industries are searching for an efficient, sustainable lubrication method for machining processes. MQL aided machining, also known as near-dry machining, is thought to be the most effective way to increase the environmental friendliness and machining performance while lowering the amount of lubrication used throughout the process (Revuru & Rao, 2017; Vishal et al., 2015). A limited amount of cutting fluid is combined with highly pressurised air and delivered to the cutting area during MQL aided machining (Revuru & Rao, 2017; Revuru et al., 2020). However, the quantity of lubricant used for different studies varies considerably. The cutting fluid consumption in MQL-assisted machining operations is less than 10–100 mL/h (Obikawa et al., 2008; Tai et al., 2014; Sharma et al., 2009). While comparing the cutting fluid consumption in conventional flood cooling (approximately 1,200 L/h), the cutting fluid usage in MQL is considerably very low (Tai et al., 2014). The machining performance under MQL settings was discovered to be quite comparable to that of conventional flood cooling (Barczak et al., 2010; Morgan et al., 2012). The MQL has drawn a lot of attention because to its low cutting fluid consumption and strong machining capabilities.

Firstly, the term *minimum quantity lubrication* is used in literature in the case of lubrication of bearing by Weck and Koch in 1993 (Weck & Koch, 1993). The research related to MQL during machining was started in 1997 for the grinding process (Brinkmeier et al., 1997) and consequently, MQL-related research became popular for all metal cutting processes. The proper lubrication, cooling, chip removal, and heat removal from the cutting zone are the main aim of cutting fluid (Pervaiz et al., 2019; Revuru & Rao, 2017). Since the primary goal of any machining process is to reduce the machining cost and improve productivity; most of the machining operations were performed at high cutting speeds. It was observed that flood cooling is very efficient at a low cutting speed. However, due to the ineffective cutting fluid penetration into the tool-workpiece contact area during high speed cutting, flood cooling is ineffective (Boswell et al., 2017). MQL is considered an alternative way to cool the cutting zone and improve the machining performance in such cases. Figure 6.1 shows the increase in various stakeholders' acceptance of

DOI: 10.1201/9781003328742-6

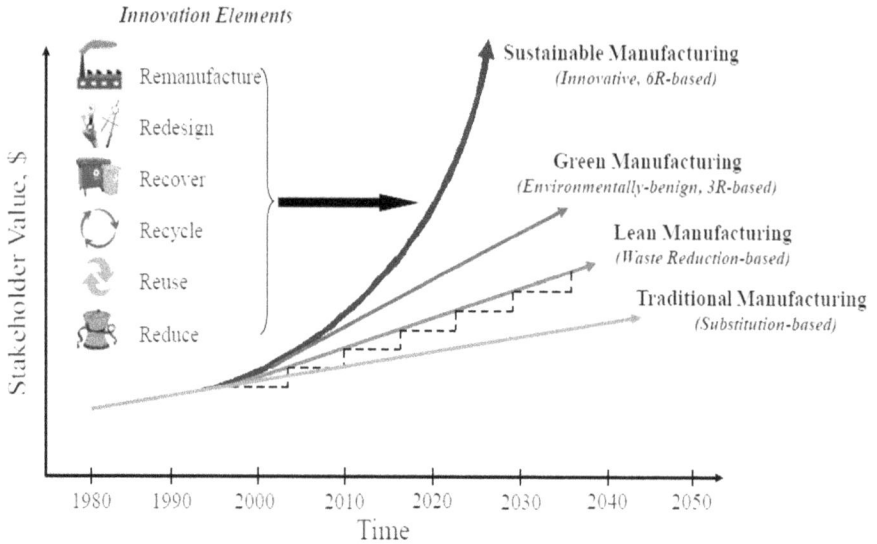

FIGURE 6.1 Progression of sustainable manufacturing (Jayal et al., 2010).

sustainable manufacturing practices (Jayal et al., 2010). However, there hasn't been much industry adoption of the MQL idea. MQL is a new technique for green manufacturing that has emerged from many sustainable production techniques. This chapter's major goal is to offer a thorough analysis of the MQL's performance in relation to several machining operations, including turning, drilling, milling, and grinding.

6.2 MQL'S IMPACT ON TURNING

One of the most popular machining processes is turning, which is mostly used to produce cylindrical components and features (Kalsi et al., 2010). The main topic of discussion in this session is how MQL affects turning performance. The session is broken up into four areas for in-depth analysis: the impact of MQL on cutting forces and temperature, tool wear, surface quality, and chip formation.

6.2.1 THE IMPACT OF MQL ON CUTTING FORCES AND TEMPERATURE

Important performance variables in a turning operation are cutting force and cutting temperature. Masoudi et al.'s (2018) investigation into the turning of AISI 1045 steel discovered that the MQL machining significantly lowered the cutting force in comparison to the dry and flood cooling. The cutting force variation for turning AISI 1045 steel is shown in Figure 6.2 for various lubrication systems. It was discovered that MQL machining uses a cutting force that is 42% lower than dry machining and 18% lower than continuous cooling. In the area of contact, a layer with low shear strength can be formed by a good cutting fluid. This aids in lowering chip adhesion and friction, which in turn lowers cutting force. Additionally, in

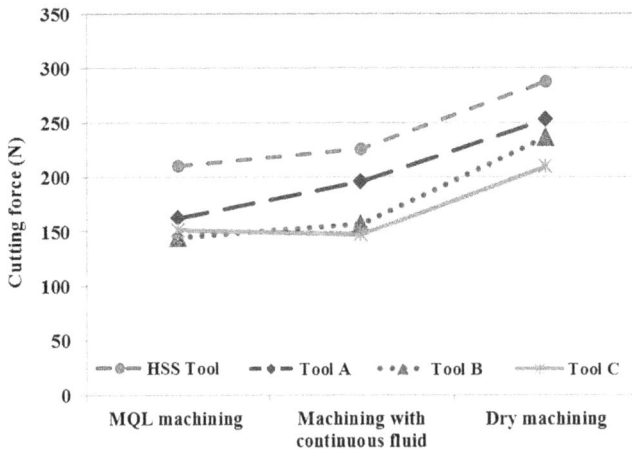

FIGURE 6.2   Changes in cutting force under different lubrication environments (Masoudi et al., 2018).

MQL machining, the oil droplets can quickly penetrate and reach the cutting zone, reducing friction greatly. The "embrittlement effect of the lubricant" works in conjunction with adhesion and friction to lessen cutting force.

The tension necessary for chip fracture is reduced by the embrittlement. Additionally, it has been demonstrated that the lubricant film prevents the micro-crack from closing as a result of the workpiece's plastic deformation (Patole & Kulkarni, 2017). Due to the stress concentrator function of these micro-cracks, only low energy is needed for machining. As a result, only MQL conditions—as opposed to dry and flood conditions—require less cutting force. Numerous studies have been done in this area since cutting parameters play a significant effect in cutting force. When turning AISI 4140 steel, Gurbuz et al. (2020) looked at the cutting force and found that it rises as the feed increases. Similar kinds of outcomes were seen throughout several works (Davim et al., 2007; Mallesha & Nayaka, 2018). The primary cause of the rise in cutting force with feed is the increase in chip cross-section area brought on by the increase in feed rate (Behera et al., 2018; Mallesha & Nayaka, 2018).

The cutting force fluctuation with feed rate during MQL-assisted turning of AISI 4140 steel is shown in Figure 6.3. When cutting speed is increased, the cutting force often lowers as well. This might be brought on by the secondary deformation zone's decreased shear strength. Due to the temperature increase caused by the rapid cutting speed, shear strength is reduced (Ciftci, 2006; Pathak et al., 2013). Because of the decreased friction at the chip-tool interface, the primary cutting force is decreased (Dhar et al., 2007; Mallesha & Nayaka, 2018; Pathak et al., 2013). The relationship between cutting force, flow rate, and cutting speed is depicted in Figure 6.4.

At high cutting velocity (150 m/min), an increase in principal cutting force was observed for certain MQL flow rates (Figure 6.4). This may be due to the fact that at very high cutting speed, the lubricant may rapidly vaporise because of high heat

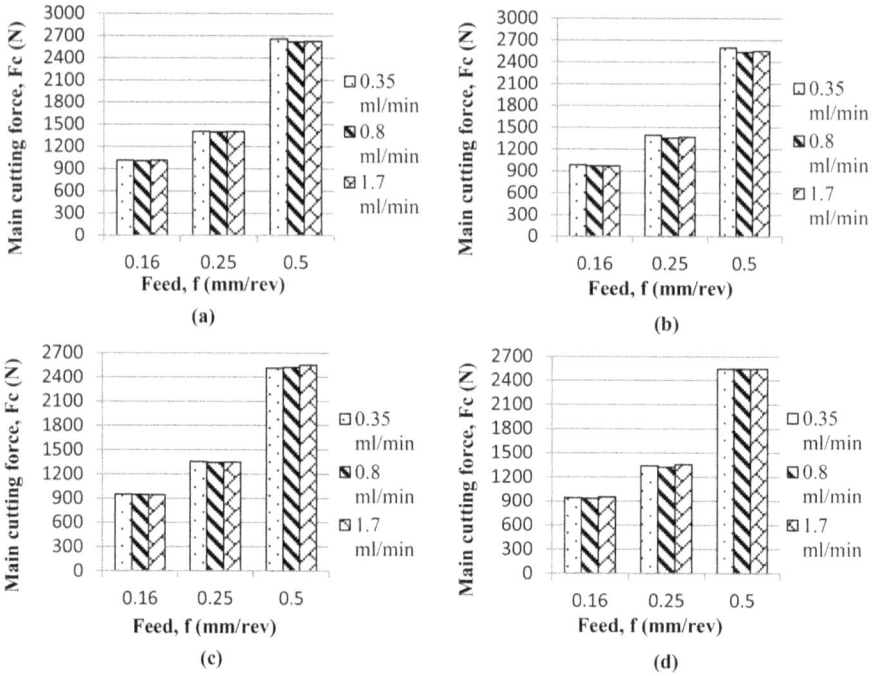

FIGURE 6.3 Effect of main cutting force with flow rates and feeds at (a) 75 m/min, (b) 100 m/min, (c) 125 m/min, (d) 150 m/min (Gurbuz et al., 2020).

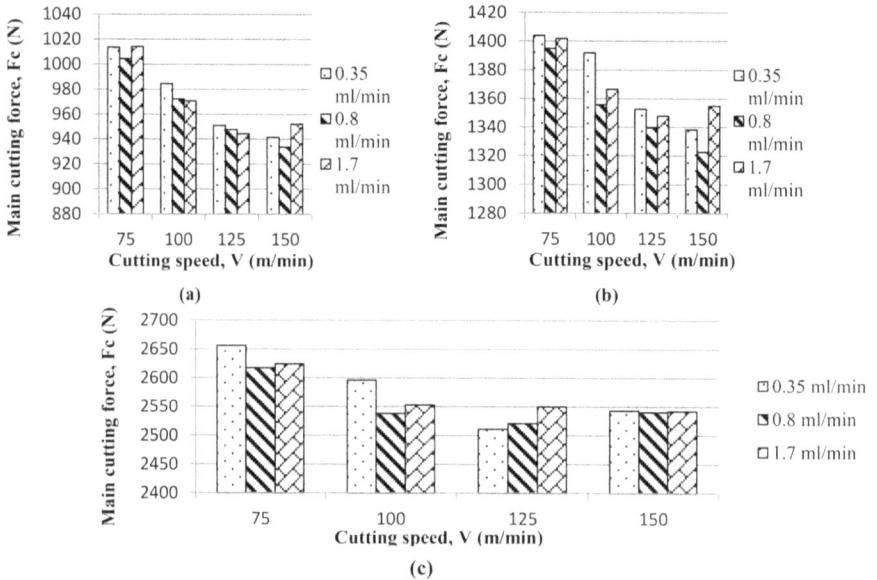

FIGURE 6.4 Effect of main cutting force with flow rate and cutting speed: (a) 0.16 mm/rev, (b) 0.25 mm/rev, (c) 0.5 mm/rev (Gurbuz et al., 2020).

generation and hence did not perform the lubricating property (Sani et al., 2019; Sani et al.,2019). The cutting force generally decreases with the flow rate (0.35 mL/min to 0.8 mL/min). But after reaching an effective MQL flow rate, cutting force increases with an increase in flow rate (0.8 mL/min to 1.7 mL/min). The effect of lubricant depends on the amount of lubricant, lubrication property, workpiece surface quality, workpiece material properties, and cutting tool properties (Sani et al., 2019; Sani et al.,2019). Hence, identifying the maximum effective MQL flow rate is highly important. When the flow rate is beyond the effective MQL flow rate, the lubricant diffuses over the cutting zone and may not change (Ji et al., 2014). Therefore, above the maximum effective MQL flow rate, the cutting force increases with the flow rate (Behera et al., 2018; Ji et al., 2014). Based on Figures 6.3 and 6.4, the maximum value of the effective MQL flow rate is 0.8 mL/min for the cutting conditions and turning of AISI 4140 steel. Similar to cutting force, cutting temperature also has a significant role in cutting performance. In a metal cutting process, the heat is developed at primary, secondary and ternary deformation zones. Out of that, the maximum temperature in a cutting becomes at the tool-chip interface. Dhar et al. (2006) performed an experimental investigation on cutting temperature using a thermocouple during turning AISI-1040 steel for wet, dry, and MQL conditions.

In Figure 6.5, the effects of dry, wet, and MQL conditions on the tool-chip interface temperature and cutting velocity (Vc) were depicted. Due to the difficulty of fluid penetration into the tool-chip interface, conventional lubricant usage is generally unable to effectively reduce the temperature at the tool-chip interface. However, the tool-chip interface in the MQL system is easily accessible to cutting fluid (Boswell et al., 2017). According to Figure 6.5, this might drastically lower the machining temperature under various cutting circumstances. Reaching the MQL jet

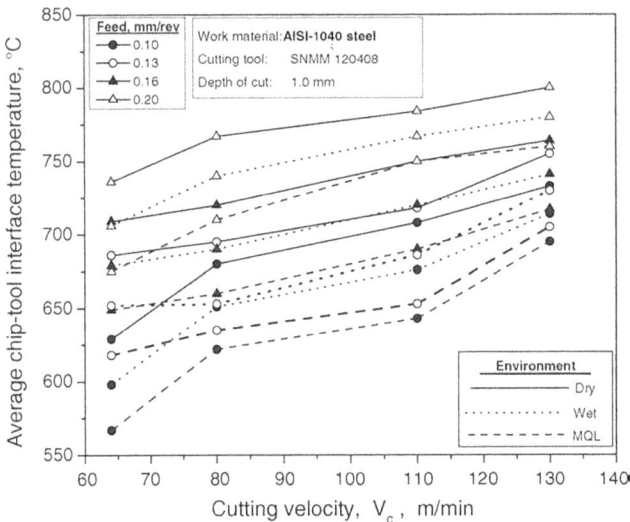

FIGURE 6.5   Effect of temperature with cutting velocity and feed rate for wet, MQL, and dry conditions during turning (Dhar et al., 2006).

closer to the tool-chip interface was made possible by the cutting edges grooves, rake surface hills, and shorter tool-chip contact length (Dhar et al., 2006). The difference in the tool-chip interaction between MQLs operating at various cutting velocities is primarily responsible for the diversity in MQL effectiveness. As a result, the temperature at the tool-chip contact in MQL conditions is lower than it is in dry and wet settings.

6.2.2 THE IMPACT OF MQL ON TOOL WEAR AND TOOL LIFE

Tool wear is an important machining performance indicator that highly influences the machined surface quality, cutting force, and tool life (Anand & Mathew, 2020). For smooth, safe, and economical machining, the tool wear must control or prevent properly. The cutting tool failure is mainly due to the following reasons:

  i. Plastic deformation of tool due to high machining temperature and pressure at the machining zone
  ii. Mechanical breakage because of large cutting force, tool vibration, insufficient toughness and strength
  iii. Gradual wear at the rake and flank face

Tools, jobs, and machine tools are all seriously harmed by the first two tool failure categories. By choosing the right tool geometry, machining parameters, and tool material, this can be avoided. Failure of a tool through gradual wear cannot be stopped; it can only be controlled. A pictorial representation of the tool wear pattern in a cutting tool is shown in Figure 6.6. The following criterion can be used to explain the cutting tool failure:

  a. Massive fracture at the cutting edge
  b. Breakage of the tool or tool tip
  c. When average flank or crater wear reaches its specified limit
  d. Excessive increase in the cutting force, tool vibration, and chatter

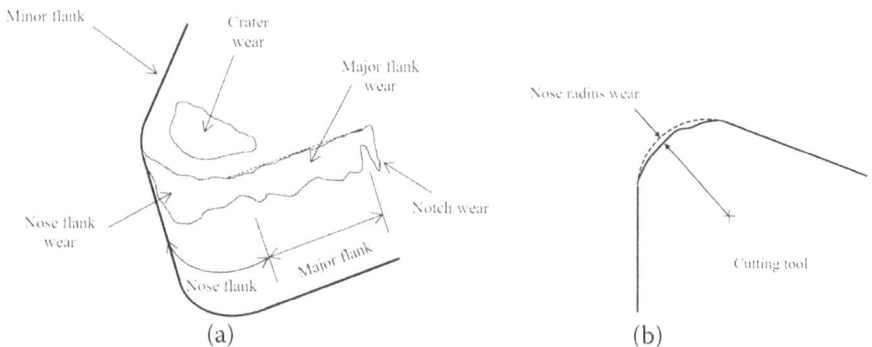

FIGURE 6.6   a) Different types of tool wear and b) nose radius wear (Derani & Ratnam, 2021).

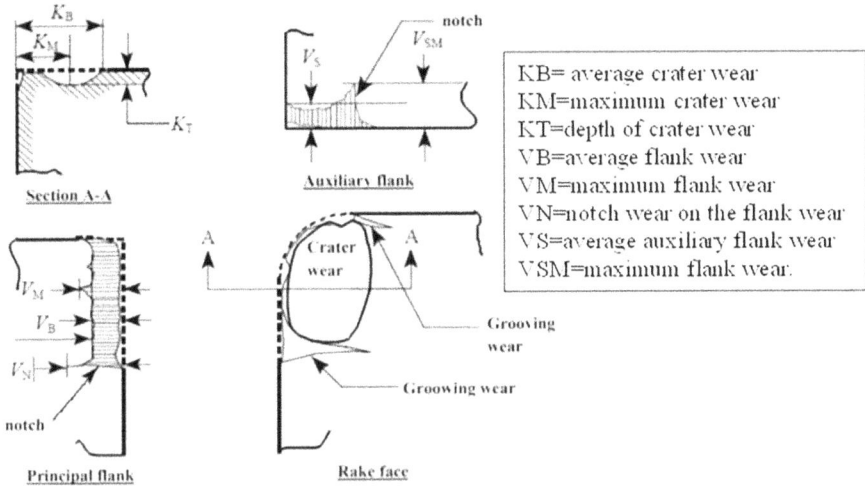

FIGURE 6.7   Wear geometry of cutting tool (Dhar et al., 2006).

e. Rapid increase in surface roughness value
f. Excessive power consumption
g. Dimensional deviation beyond tolerance
h. Adverse chip formation

During the turning process, the cutting tool wear mainly occurred due to adhesion, chemical erosion, abrasion diffusion, and galvanic effect (Dhar et al., 2006). In the first stage of cutting, the tool wear starts at a moderately fast rate because of microchipping and fracture wear at sharp cutting edges. Figure 6.7 shows the schematic representation of tool wear for the turning tool. Out of different tool wear terminologies, the average flank wear (VB) is the most commonly used tool wear indicator due to its influence on cutting force. Generally, the term *tool life* is associated with the actual cutting time when VB approaches a critical value, which is 0.3 mm. Hence, without reducing the machining rate, it is highly required to reduce the VB value to achieve high tool life.

To enhance the tool life, many researchers successfully used MQL. Figure 6.8 displays the variation of VB with chip removal volume for different flow rates during MQL-assisted turning of AISI 4140 (Gurbuz et al., 2020). From Figure 6.8, it is clear that the value of VB increases with chip removal volume for all flow rates. From Figure 6.8, it is clear that the VB increases with a decrease in MQL flow rate during machining AISI 4140. This is due to the fact that the increase in flow rate reduces the coefficient of friction between the tool-chip interface (Sarikaya & Gullu, 2015).

Figure 6.9 shows the cutting tool images under different MQL flow rates. These SEM and optical images can be used to understand the tool wear behavior in detail. From Figure 6.9, the flank wear is clearly visible due to the abrasion scratches for all flow conditions. These abrasion marks were observed due to the abrasion of

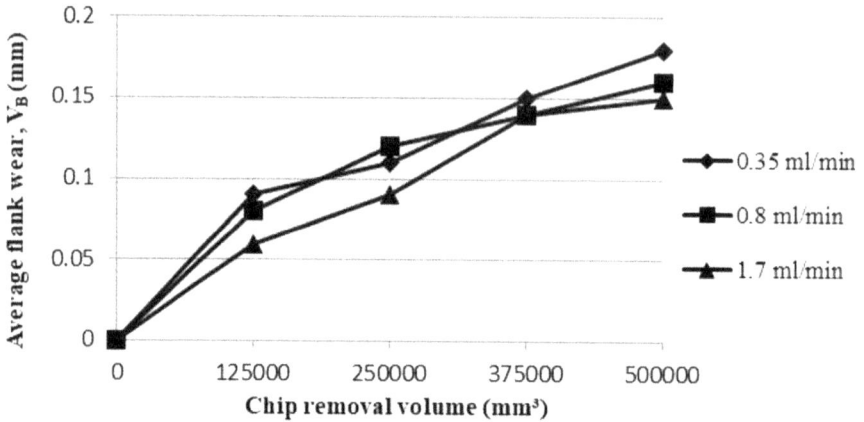

FIGURE 6.8 Flank wear variation with chip removal volume and MQL flow rate (V = 125 m/min, f = 0.16 mm/rev, a = 2.5 mm) (Gurbuz et al., 2020).

workpiece hard particles on the cutting tool. These marks are observed as parallel grooves in the machining path. For all cutting conditions, adhesive wear was also observed due to the sticking of workpiece material on the crater surface. The workpiece adhesion decreases with an increase in flow rate due to the decrease in heat and friction in the machining zone (Sarikaya & Gullu, 2015).

6.2.3 THE IMPACT OF MQL ON SURFACE ROUGHNESS

Since the poor surface roughness of the machined components/features leads to improper fitting and dimensional inaccuracy of the parts, it is important to study and improve the surface quality of the machined features. As the quality of the machined surface is very important for most engineering applications, many researchers studied the influence of MQL on surface quality. The surface roughness was reduced significantly by the application of MQL in comparison with dry cutting. Gurbuz et al. (Gurbuz et al., 2020) studied the MQL-assisted turning of AISI 4140 steel and found that the surface roughness increases with feed rate (Gurbuz et al., 2020; Sarikaya, M. & Gullu, 2014). Equation 6.1 shows that the average surface roughness is inversely proportional to nose radius (r) and directly proportional to square of feed rate (f) (Gurbuz et al., 2020).

$$Ra = 0321 f^2 / r \qquad (6.1)$$

Figure 6.10 shows the variation of surface roughness with cutting speed for different feed rates. In general, the surface roughness decreases with an increase in cutting speed (Gurbuz et al., 2020; Hwang & Lee, 2010). At a high cutting speed, the heat developed may increase and lead to plastic deformation and this result to the decrease in surface roughness value (Sarikaya, M. & Gullu, 2015). As shown in Figure 6.10, the surface roughness does not always decrease with an increase in

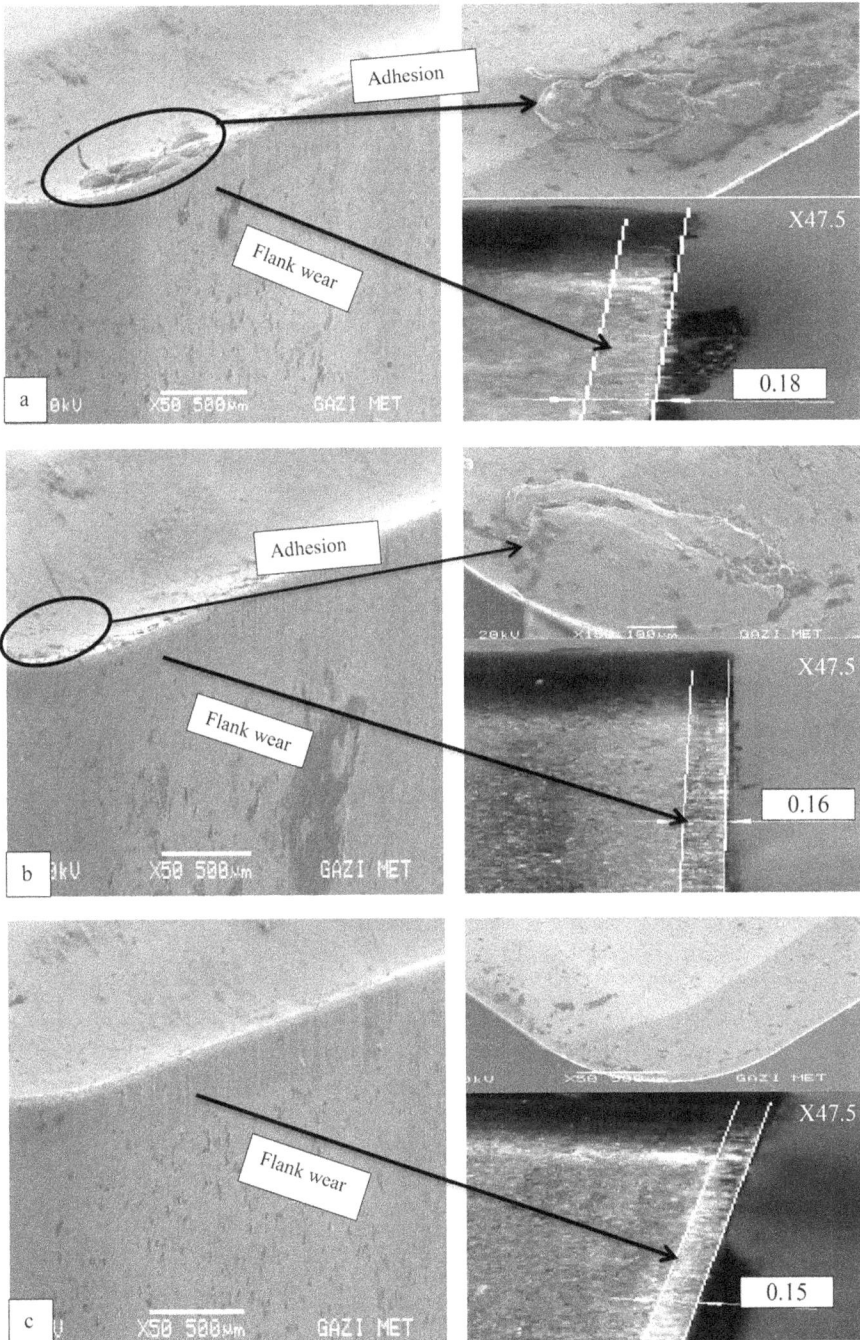

FIGURE 6.9   Cutting edges at (a) 0.35 mL/min, (b) 0.8 mL/min, (c) 1.7 mL/min (Gurbuz et al., 2020).

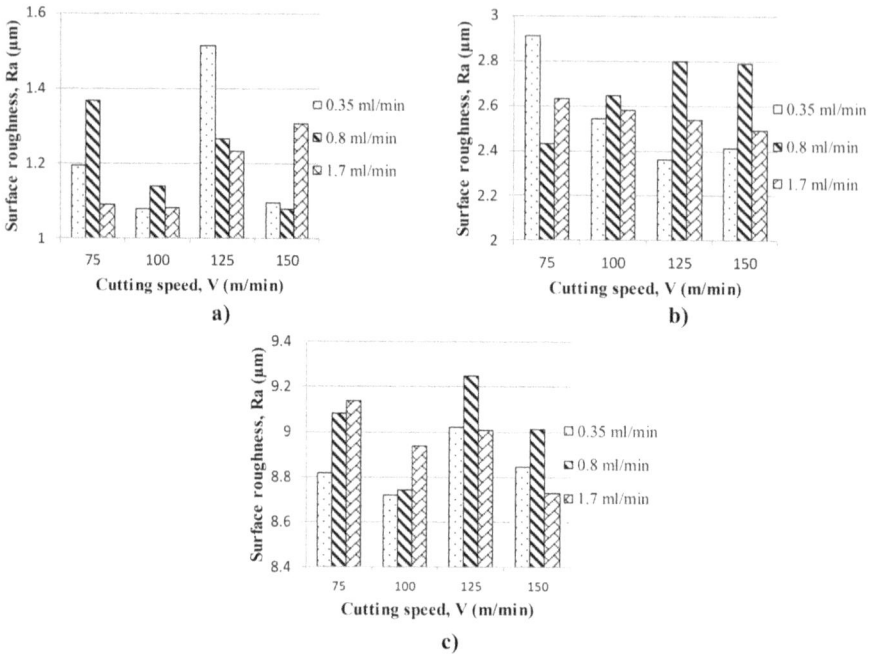

FIGURE 6.10   Effect of surface roughness with flow rates and cutting speed at (a) 0.16 mm/rev, (b) 0.25 mm/rev, (c) 0.5 mm/rev (Gurbuz et al., 2020).

cutting speed. The continuous chip formation at higher cutting speed leads to surface deformation in certain cases and this may lead to irregular variations in surface quality (Cakir et al., 2016).

From Figure 6.10, the high surface roughness was observed at 0.5 mm/rev feed and 125 m/min speed, and low surface roughness was observed at 0.16 mm/rev feed and 150 m/min cutting speed. In general, surface roughness decreases with an increase in flow rate. But, in certain cases, the long unwanted chips may wrap around the workpiece (Gurbuz et al., 2020). This may prevent the flow of lubricant toward the cutting tool-workpiece interface. As a result, an unexpected increase and decreases in surface roughness were observed (Diniz et al., 2003; Gurbuz et al., 2020). Figure 6.11 shows the optical surface topography of turned surface under dry, wet, and MQL environments. From this image, it is clear that MQL significantly enhanced the surface quality compared to dry and flood conditions.

6.2.4   THE IMPACT OF MQL ON CHIP FORMATION

The chip formation during machining is an important indication of machinability and machining performance. The chip morphology highly depends on workpiece material, cutting tool, cutting condition, and machining environment. Dhar et al. (2006) investigated the influence of cutting environments, such as dry, wet, and MQL conditions, on chip formation during turning AISI 1040 steel. In general, both

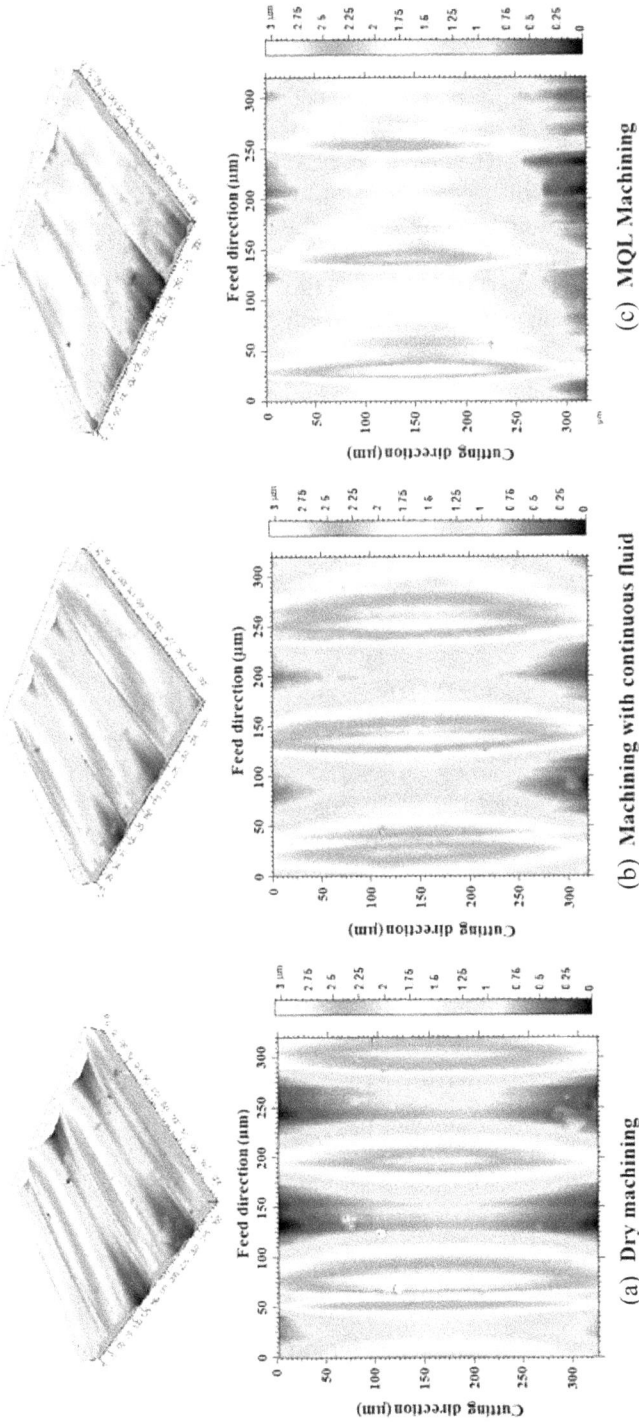

FIGURE 6.11 Topography of machined surface for flood, dry, and MQL conditions (Masoudi et al., 2018).

dry and wet machining of steel produces spiral chips with blue color. The geometry of the cutting tool significantly influences the chip formation. The insert geometry must be such that the chip initially becomes curled along the normal plane and then hits the principal flank. The chips produced under MQL conditions appeared smooth and bright. This is mainly due to the reduction in cutting temperature and friction between the tool and chip. The chip color becomes much lighter depending on the cutting velocity. Figure 6.12 shows the actual chips produced while turning steel at the feed of 0.16 mm/rev and cutting velocity 110 m/min under wet, dry, and MQL environments. The application of MQL decreases the chip reduction coefficient at low cutting speeds (Dhar et al., 2006). Davim et al. (2007) also observed that the chip formation under the MQL condition is similar to that of flood lubrication conditions.

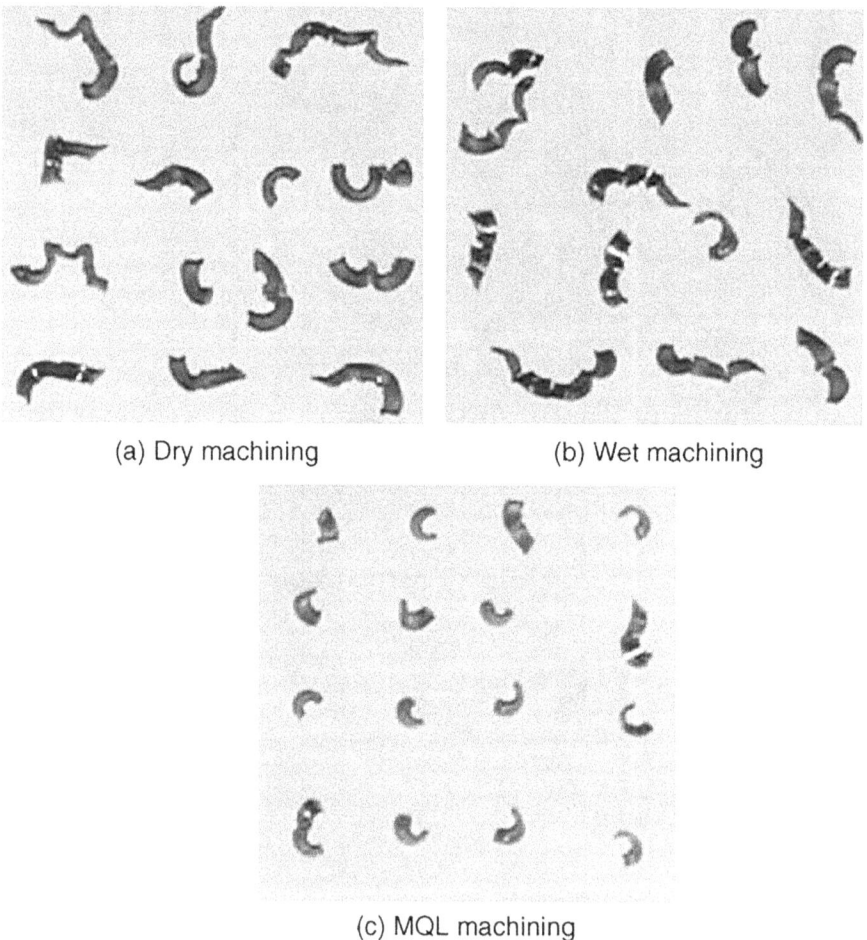

(a) Dry machining　　　　　　　　(b) Wet machining

(c) MQL machining

FIGURE 6.12　Chips formed during turning steel at 0.16 mm/rev feed and 110 m/min cutting velocity for (a) dry, (b) wet, and (c) MQL environments (Dhar et al., 2006).

6.3   EFFECT OF MQL ON MACHINING PERFORMANCES DURING MILLING OPERATION

In this section, the effect of MQL on machining performance, such as cutting force, temperature, tool wear, surface roughness, and chip morphology during milling operation, was discussed. The improvement in machining performance and the reasons for the improvement were emphasized in detail.

6.3.1   Effect of MQL on cutting forces and temperature

Narayanan et al. (2021) studied the cutting force and temperature during MQL-assisted end milling of Ti6Al4V. The milling experiments were carried out with a constant depth of cut (0.5 mm) and flow rate for both flood (flow rate = 165 L/h) and MQL (flow rate = 350 mL/h) conditions. Figure 6.13 shows the effect of axial and thrust force with feed rate for dry, flood, and MQL conditions.

It was observed that the cutting force was reduced up to 35% by the application of MQL compared to flood environments. From this figure, it can be noticed that for both flood and MQL conditions, the cutting force increases with the feed rate for all cutting speeds. This is due to the high material removal at high feed rates. As a result, thick chips were formed and resulted in the reduction in shear angle. Hence, the shear force increases and leads to an increase in primary cutting force. However, the increase in cutting force with feed rate in MQL condition is much less than that of a flood condition. In general, the value of cutting force highly depends on thermal softening and strain hardening. Initially, the cutting force increases with an increase in cutting speed due to the severe strain hardening. However, a further rise in cutting speed leads to the development of high cutting temperatures. This may lead to thermal softening and a reduction in cutting force at high speed (Hou et al., 2014; Rajaguru & Arunachalam, 2020). The reduction in cutting forces at a high cutting speed is due to improved cutting stability (Do & Hsu, 2016). Similar to cutting force, cutting temperature also shows a similar trend with feed rate. The variation in machining temperature with feed rate and cutting speed for both flood and MQL is shown in Figure 6.14.

From Figure 6.14, it was found that the cutting temperature was significantly reduced up to 47% for MQL compared to flood conditions while end milling of Ti6Al4V. The small value of cutting force and cutting temperature at a low feed rate is mainly due to the low plastic deformation and easy penetration of cutting fluid into the cutting zone (Sun et al., 2015). The small cutting force due to the effective lubrication leads to a reduction in tool wear in MQL conditions compared to flood conditions (Schoop et al., 2017). The variation of cutting force with cutting speed for MQL-assisted milling also follows a trend similar to flood conditions (Benjamin et al., 2018; Rajaguru & Arunachalam, 2020). The MQL-assisted cutting performance decreases at a high cutting speed. However, the high cutting temperature at a high cutting speed (150 m/min) leads to thermal softening and results in a reduction in cutting force. It was observed that the relationship between cutting force and speed shows a quadratic nature, even though it is highly prominent for a 0.075 mm/rev feed rate. This might be due to the dominant strain hardening over

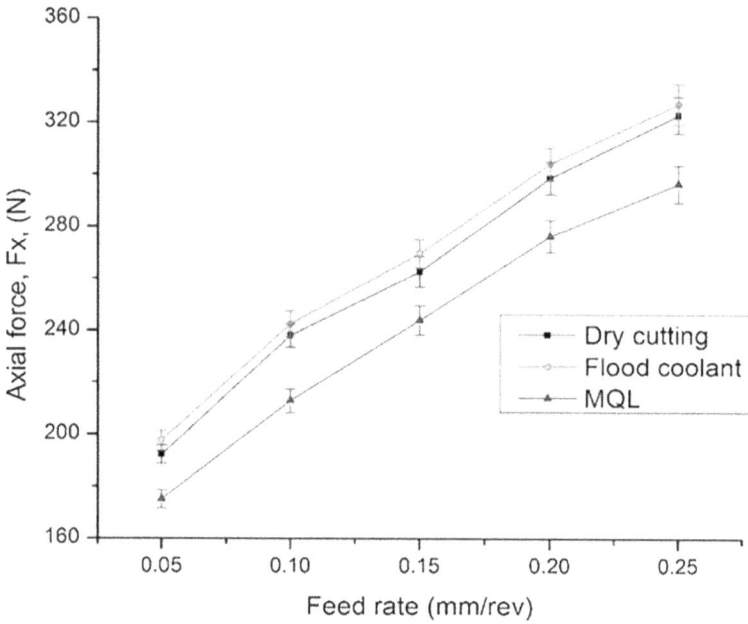

(a) Axial force

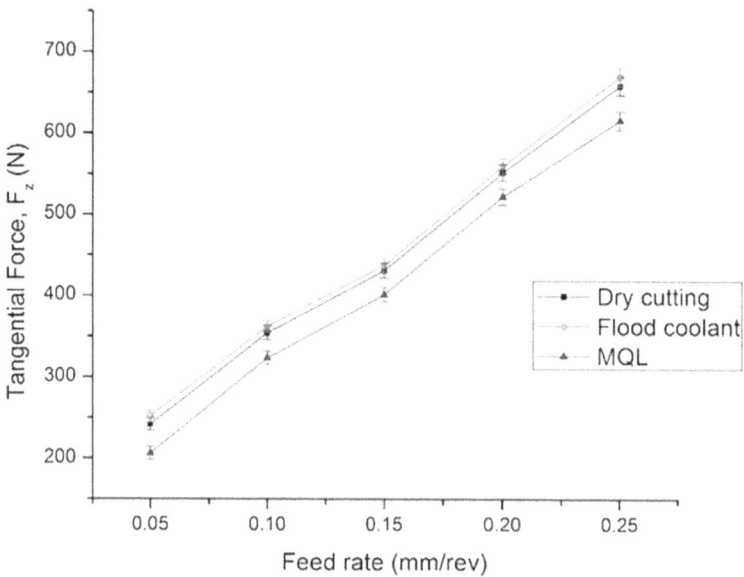

(b) Tangential force

FIGURE 6.13 Effect of cutting force with feed rate for (a) dry, (b) flood, and (c) MQL conditions (Rajaguru & Arunachalam, 2020).

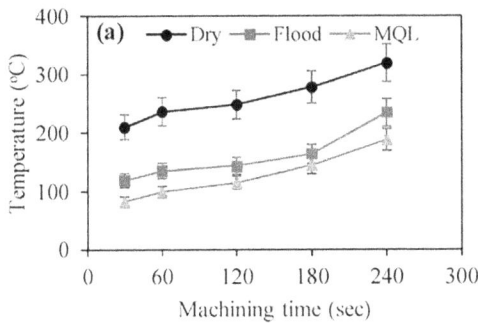

FIGURE 6.14   Effect of cutting temperature with machining time for (a) dry, (b) flood, and (c) MQL conditions (Khan & Maity, 2018).

thermal softening at 120 m/min speed and 0.075 mm/rev feed. It was found that the feed rate has the highest influence on strain hardening (Bordin et al., 2014; Pramanik & Littlefair, 2015) and frequency shear banding (Pramanik & Littlefair, 2015). The high cutting temperature and low cutting force at 150 m/min and 0.075 mm/rev suggest that thermal softening balances strain hardening.

During a machining process, the cutting parameters and tool geometry directly influence temperature generation and heat dissipation. In a cutting process, a huge amount of the mechanical energy from the machine tool is converted to heat. This heat is dissipated through the chip, workpiece, and cutting tool. In many cases, especially during dry machining, overheating creates several issues. Therefore, it's highly important to use suitable cooling strategies to reduce cutting temperature and improve the machining performance (Anand & Mathew, 2021). Even though dry machining leads to severe tool wear, poor tool life, huge cutting temperatures, and poor surface quality, it is conventionally preferred for machinability studies. The MQL application is an improved strategy to overcome these issues in an environmentally friendly way due to its thermal and tribological behavior (Anand & Mathew, 2021). Figure 6.15 shows the variation of cutting temperature with feed rate for both dry and MQL conditions (Ji et al., 2014). From this figure, it was clear that irrespective of machining parameters, the MQL machining significantly reduced the cutting temperatures.

6.3.2   EFFECT OF MQL ON TOOL WEAR

Similar to the turning process, tool wear in the milling process is also an important performance parameter, which significantly influences the machined surface quality and cutting mechanism (Anand & Mathew, 2020; Rajaguru & Arunachalam, 2020). In most milling operations, flank wear is considered the major indicator of tool life. The increases in flank wear result in the development of high cutting pressure and temperature. To reduce tool wear in an eco-friendly way, many researchers discussed the effect of MQL on wear behavior. Salur et al. (2021) studied the effect of MQL on tool wear during the milling of AISI 1040 steel. It was found that, as cutting progresses, the cutting tool loses its main cutting edge, which leads to severe cutting

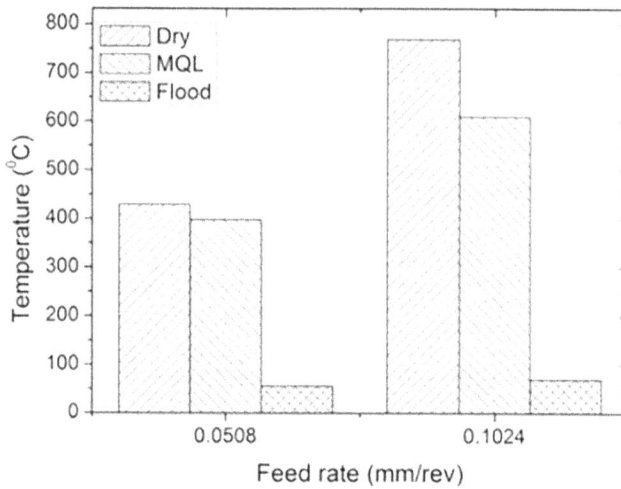

FIGURE 6.15   Effects of the cutting environment on the cutting temperature (Ji et al., 2014).

force, poor surface quality, chatter vibration, and power consumption. Therefore, the tool wear analysis is highly important. In general, the mechanical pressure on the milling cutter is a key factor which triggers the abrasion between the chip and tool (Hou et al., 2014). During the milling operation, both the cutting parameter and the milling mechanism have a significant role in the wear behavior. The increase in the level of cutting parameters leads to an increase in cutting load and which results in faster tool wear.

Figure 6.16 shows variation in wear for different conditions during the milling of In718. From these figures, a significant reduction in tool wear was observed for MQL compared to dry conditions. Maximum flank wear (VB_{max}) reduction of up to 30.9% was observed for MQL compared to dry conditions. This is mainly due to the improved tribological performance by proper lubrication and cooling of MQL from the jest of pressurized air and oil droplets (Anand & Mathew, 2021).

The cutting speed significantly influences the tool flank wear (Salur et al., 2021). As both cutting speed (100 m/min to 150 m/min) and feed rate (0.1 mm/rev to 0.15 mm/rev) increased, the flank wear was also observed to be increasing due to the hammering effect of the milling mechanism.

Figure 6.17 shows the adhesion on the tool under dry cutting (air), MQL (mist) and flood cooling (coolant) during end milling of Ti6Al4V. From Figure 6.17(a), the workpiece adhesion was found to be severe on the rake face in dry cutting of Ti6Al4V. As shown in Figure 6.17(b) and 6.17(c), the application of cutting fluid significantly reduces the adhesion at the tool workpiece and tool chip interface. This is due to the fact that the cutting fluid can carry chips from the cutting zone to reduce the interaction between the hot chip and the milled surface.

Figure 6.18 shows the variation of tool life with cutting speed during end milling of Ti6Al4V. For the same cutting speed, the tool life observed under MQL conditions (maximum tool life = 800 min) is very much higher than that of dry and

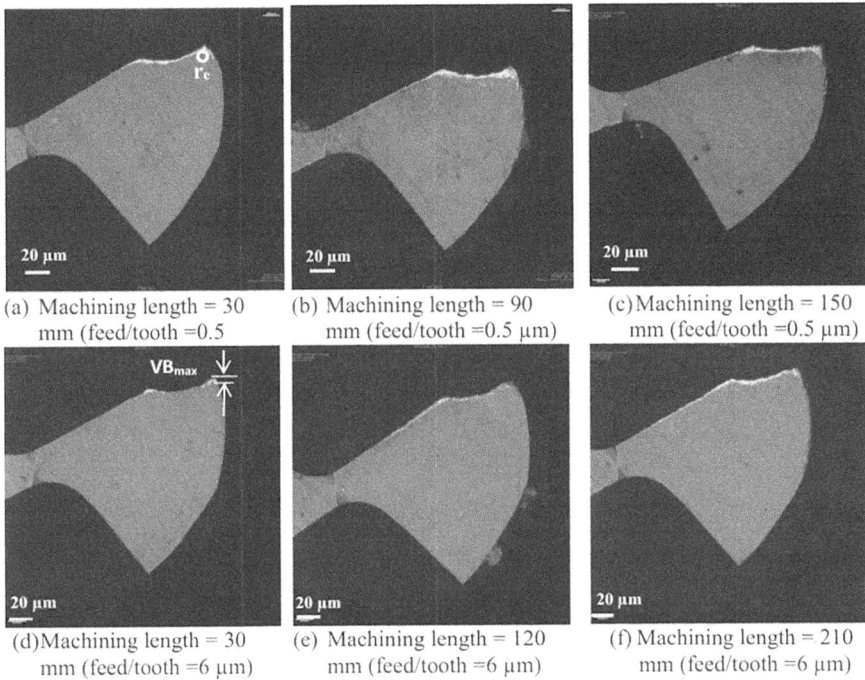

(a) Machining length = 30 mm (feed/tooth =0.5

(b) Machining length = 90 mm (feed/tooth =0.5 µm)

(c)Machining length = 150 mm (feed/tooth =0.5 µm)

(d)Machining length = 30 mm (feed/tooth =6 µm)

(e) Machining length = 120 mm (feed/tooth =6 µm)

(f) Machining length = 210 mm (feed/tooth =6 µm)

FIGURE 6.16 Progressive tool wear for different conditions during milling of Inconel718 (Anand & Mathew, 2021).

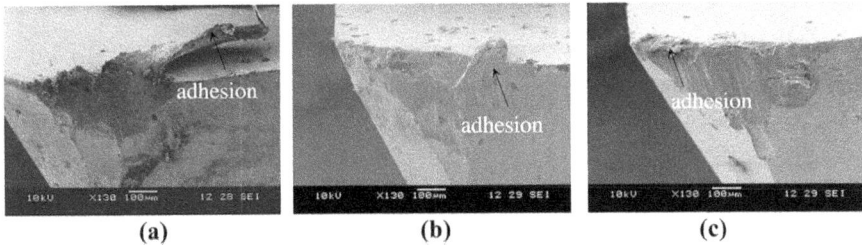

FIGURE 6.17 Adhesion during endmilling of Ti6Al4V for (a) dry cutting, (b) MQL, and (c) flood conditions (Sun et al., 2006).

flood conditions. It was also observed that flood cooling can be efficiently adapted to low-speed titanium machining only. In certain conditions, the tool life under dry conditions exceeds that of flood conditions. This is due to the solid lubrication effect and oxidation of TiC and carbide tools at dry cutting temperatures (Rahman et al., 2002; Sun et al., 2006).

Figure 6.19 shows the SEM micrographs of cutting tools at different machining conditions. The main tool failure modes in titanium machining are observed to be crater wear, flank wear, catastrophic tool failure, flaking, and chipping. The catastrophic failure observed in flood cooling is mainly due to the high thermal stress

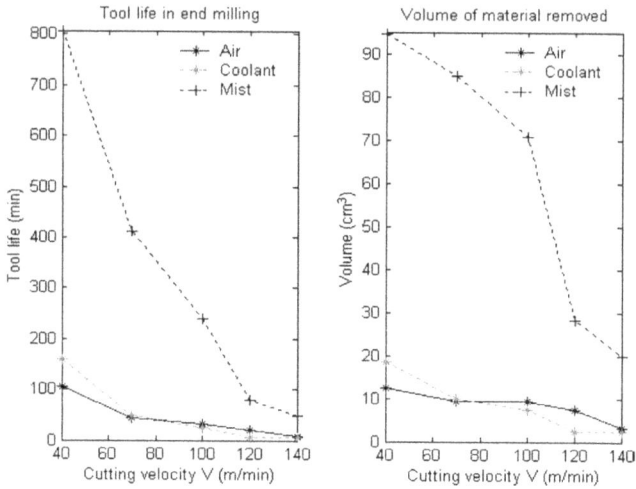

FIGURE 6.18 Variation of tool life with cutting speed for flood cooling (coolant) dry cutting (air), and MQL (mist) conditions during end milling of Ti6Al4V (Sun et al., 2006).

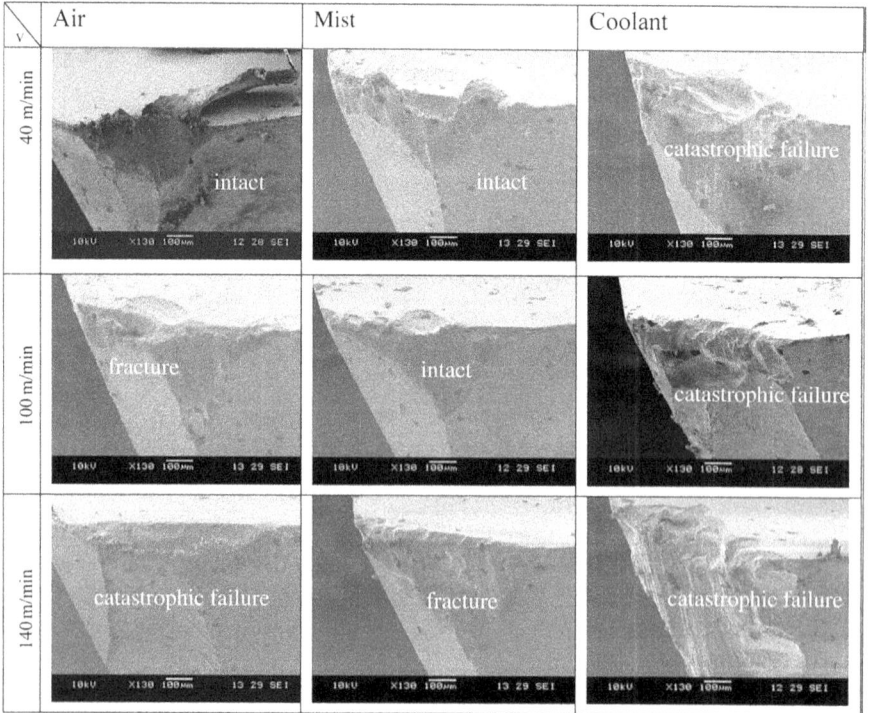

FIGURE 6.19 SEM images cutting tool at different cutting speeds in MQL (mist), flood cooling (coolant), and dry cutting (air) during end milling of Ti6Al4V (Sun et al., 2006).

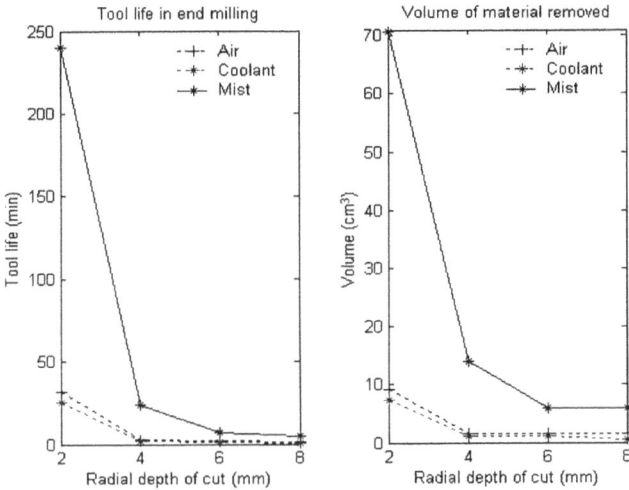

FIGURE 6.20 Variation of tool life with depth of cut for MQL (mist), flood (coolant), and dry cutting (air) during endmilling of Ti6Al4V (Sun et al., 2006)

caused by the mass of flood coolant. However, under MQL conditions, tool wear was observed to be less due to efficient lubrication, low thermal stress, and formation of TiC (Sun et al., 2006). The high cutting speed leads to severe plastic deformation and thermal cracks and leads to catastrophic failure. Due to this reason, the edge deformation and flank wear increase with cutting speed.

Figure 6.20 shows the variation of tool life with radial depth of cut for end milling of Ti6Al4V. It was found that the MQL machining tool shows the highest tool life and MRR for the same radial depth of cut. As the depth of cut increases, the tool life becomes small for all cooling conditions due to the increase in machining temperature, plastic deformation and thermal crack. For the above 4 mm radial depth of cut, not much variation in tool life between flood and dry cutting was observed. This is due to the formation of friction minimizing agents such as TiC in dry cutting and which provides an effective lubrication in the tool workpiece interface.

Figure 6.21 displayed the variation of tool life with feed rate and observed that the tool life decreases with increasing feed rate for dry, flood, and MQL conditions. MQL application significantly reduced the tool life compared to other conditions. For 0.05 mm/tooth, the tool life was observed to be 136.9 min for MQL condition, 77.8 min for dry cutting, and 77 min for flood cooling. As feed rates increase, more workpiece material will be removed and which increases the cutting load. The tool failure is observed to be more prominent at a high feed rate for flood cooling than in MQL conditions (Sun et al., 2006).

6.3.3 EFFECT OF MQL ON SURFACE ROUGHNESS

The machined surface quality is an important performance parameter which significantly influences the mechanical strength of components (Da Silva et al., 2007).

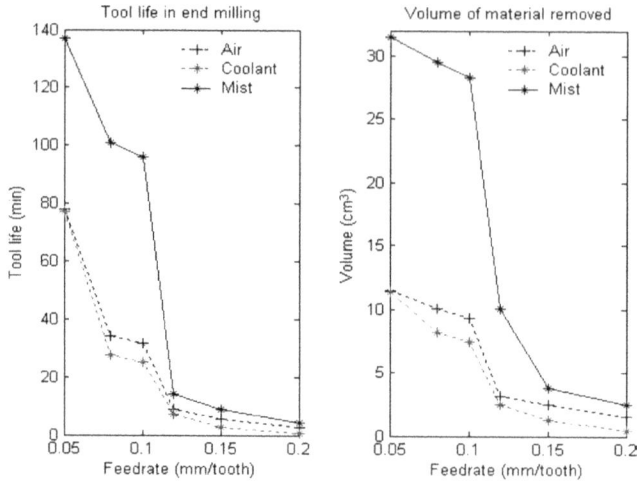

FIGURE 6.21   Variation of tool life with feed rate while end milling of Ti6Al4V for dry, flood, and MQL conditions (Sun et al., 2006).

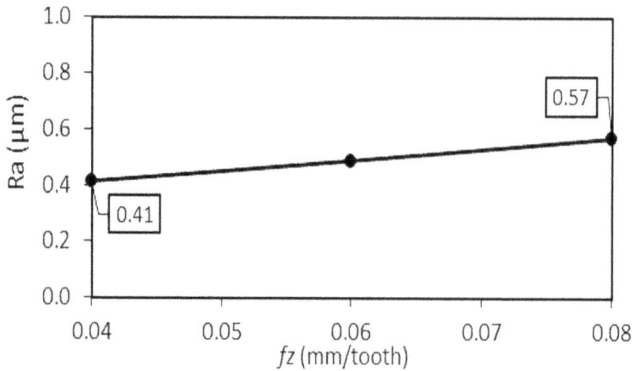

FIGURE 6.22   Effect of Ra with feed during MQL-assisted milling of Ti6Al4V (Hassanpour et al., 2020).

Hassanpour et al. (2020) investigated the effect of MQL on surface roughness during milling of Ti6Al4V. Figure 6.22 shows the effect of surface roughness during MQL-assisted high-speed milling (HSM) of Ti6Al4V. It was observed that the increase in feed rate led to an increase in surface roughness up to 34%. But the increase in feed rate results to high chip load and which results to an increase in tool vibration, cutting force and surface roughness. Figure 6.22 displays the effect of surface roughness with cutting speed for a different feed rate during MQL-assisted milling of Ti6Al4V.

From Figure 6.23, it was clear that the roughness decreases significantly with cutting speed due to the easy plastic deformation and thermal softening of workpiece materials at high temperatures. This may lead to a reduction in cutting forces and friction. In general, low cutting force makes the process more stable and results

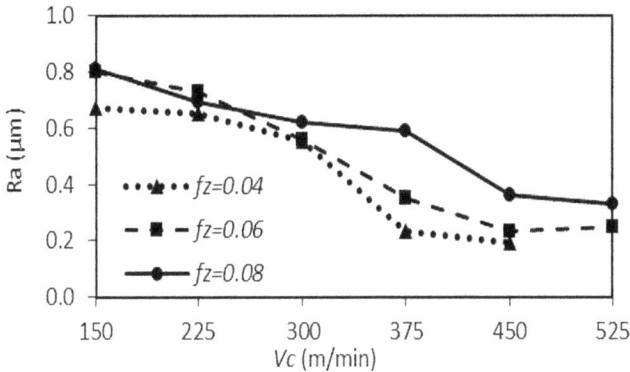

FIGURE 6.23 Effect of Ra with feed/tooth and cutting speed during MQL-assisted milling of Ti6Al4V (Hassanpour et al., 2020).

in an improvement in surface quality. For 0.06 μm/tooth, the minimum surface roughness was observed at a cutting speed of 450 m/min, which is 63% less than compared to 150 m/min cutting speed. In MQL-assisted machining, the high-pressure oil droplets are properly sprayed into the machining zone. This improves the lubricating performance and machined surface quality (Anand & Mathew, 2020). But after increasing the cutting speed above a critical cutting speed of 450 m/min, the influence of cutting speed on surface roughness was found to be less. It was also observed that at a low feed rate and 300 m/min cutting speed, the surface roughness decreases rapidly. This speed is called as the threshold speed to start the HSM process. But, at high feed rate conditions, this rapid change occurs at 375 m/min cutting speed. Figure 6.24 shows the machined surface topography while milling Ti6Al4V. At lower cutting speeds, a large number of surface smears, cavities, and tears were observed. As the cutting speed increases, these defects were observed to be very less.

6.3.4 EFFECT OF MQL ON CHIP MORPHOLOGY

In a machining process, the chip formation can give a lot of information, such as machined surface quality, tool wear, cutting force, etc. Rahman et al. (2002) investigated the chip formation in MQL-assisted milling of steel. Figure 6.25 shows the images of chip shapes and the detailed SEM micrographs for dry, flood, and MQL-assisted milling of steel. From this figure, not much difference in the length of the chip was obtained for flood cooling. However, the high thermal stress due to the high coolant flow in flood conditions leads to the fracture of the serrated chip, as shown in Figure 6.25b. This shows that the temperature was lower. But there is no evidence of high thermal stress in MQL conditions, as shown in Figure 6.25c. This is because of the higher specific heat capacity of water (4.18 kJ/kg K) than that of air (1.08 kJ/kg K). Similarly, dry milling chips are observed to have high serrations due to the severe shearing during dry cutting, as shown in Figure 6.25a.

(a)

(b)

(c)

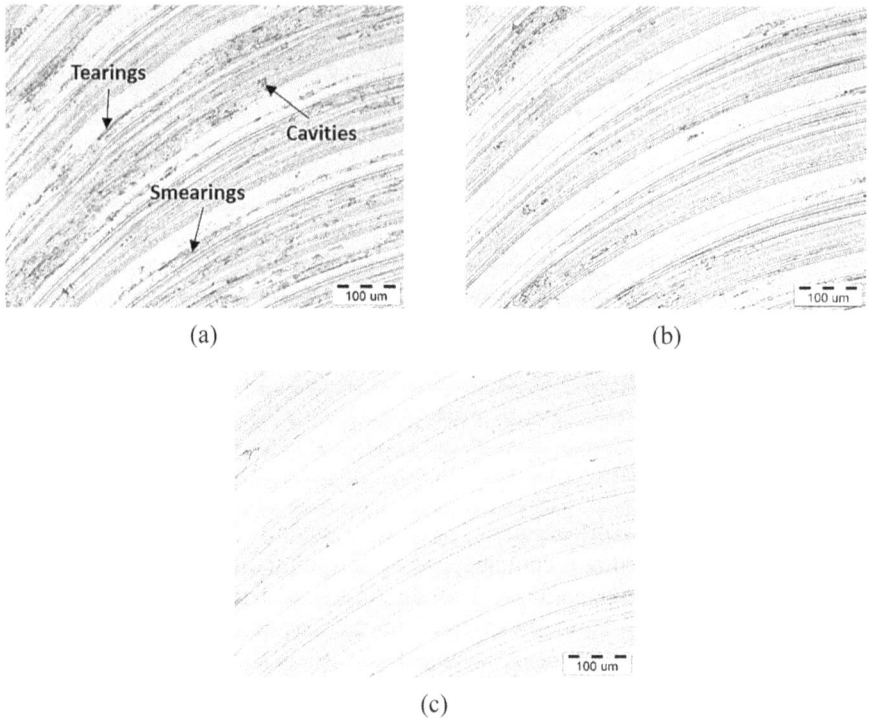

FIGURE 6.24 Surface topography of MQL-assisted milled Ti6Al4V surfaces for 0.04 mm/tooth at (a) 150 m/min, (b) 300 m/min, and (c) 450 m/min (Hassanpour et al., 2020).

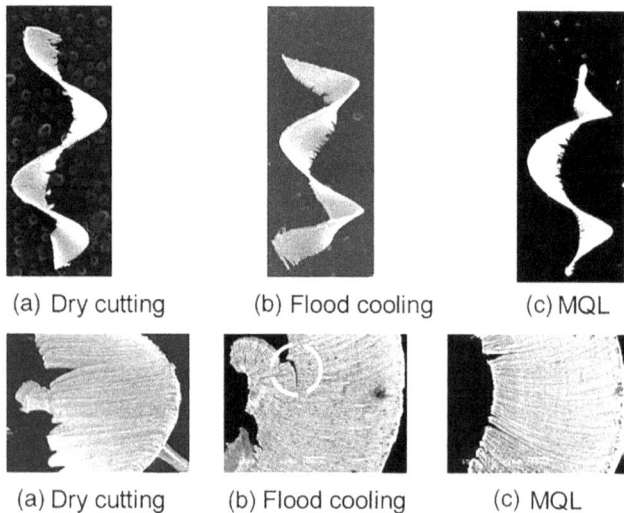

(a) Dry cutting (b) Flood cooling (c) MQL

(a) Dry cutting (b) Flood cooling (c) MQL

FIGURE 6.25 Images of chips under dry, flood, and MQL conditions (Rahman et al., 2002).

6.4 EFFECT OF MQL ON MACHINING PERFORMANCES DURING DRILLING

In this session, the effect of MQL on machining performance, such as thrust force, toque, tool wear, surface quality, and chip formation during drilling, was presented.

6.4.1 EFFECT OF MQL ON THRUST FORCE AND TORQUE

In a drilling operation, thrust force and torque are important factors that determine the cutting performance. Figure 6.26 shows the effect of cutting environments on thrust force and force during the drilling of austempered ductile iron (ADI). It was found the tool life in MQL-assisted drilling was significantly improved. The tool life of MQL-assisted drilling was 110 holes, and for dry drilling tool life was 80 holes (Meena & El Mansori, 2011). Whereas in flood conditions, the tool can drill more than 110 holes without failure. From Figure 6.26a, the reduction in maximum average torque under MQL conditions in comparison with dry drilling was observed as 6%. For flood cooling, the reduction in maximum average torque compared to dry cutting was 24%. Similarly, the difference between the maximum average thrust force under flood and MQL conditions compared to dry drilling was 13% and 10%, respectively. Hence a considerable reduction in thrust force and torque was obtained for MQL-assisted and flood conditions compared to dry drilling.

Since both thrust force and torque are significantly reduced for MQL conditions compared to dry drilling, the temperature developed during drilling may be less. Figure 6.27 shows the variation of temperature with drill depth for MQL and dry environments in machining titanium aluminide. This indicates sufficient cooling and lubrication in MQL drilling compared to dry.

6.4.2 EFFECT OF MQL ON TOOL WEAR OR DRILL WEAR

Since drilling is a highly complicated oblique cutting process, the tool wear analysis becomes challenging. Figure 6.28 shows the tool flank wear variation with drill time

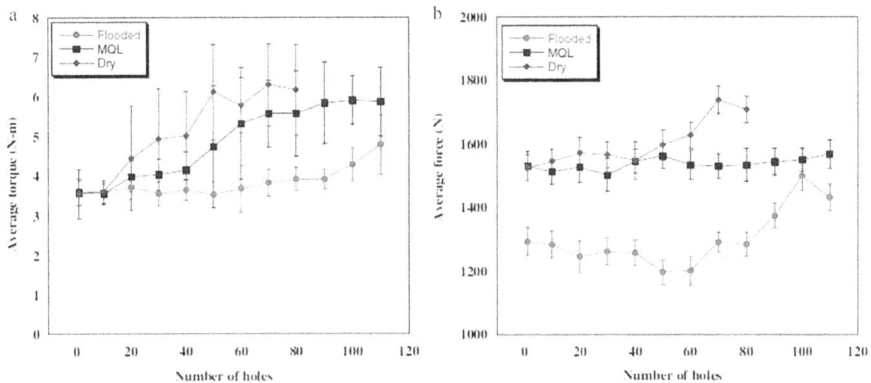

FIGURE 6.26 Comparisons of (a) torque, (b) thrust force during drilling of austempered ductile iron under flooded, MQL, and dry conditions (Meena & El Mansori, 2011).

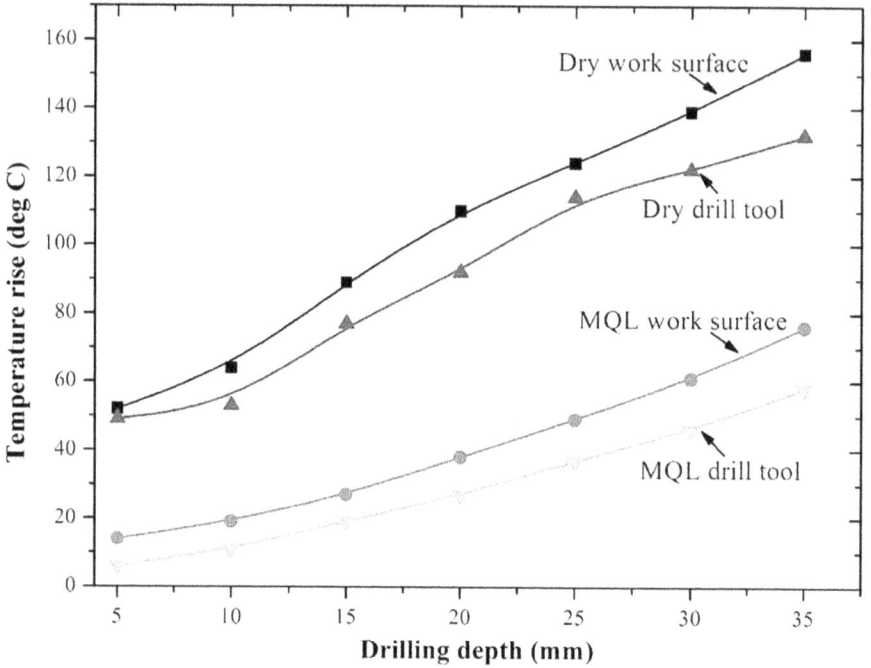

FIGURE 6.27 Drilling temperature rise under dry and MQL conditions (Nithin & Vijayaraghavan, 2018).

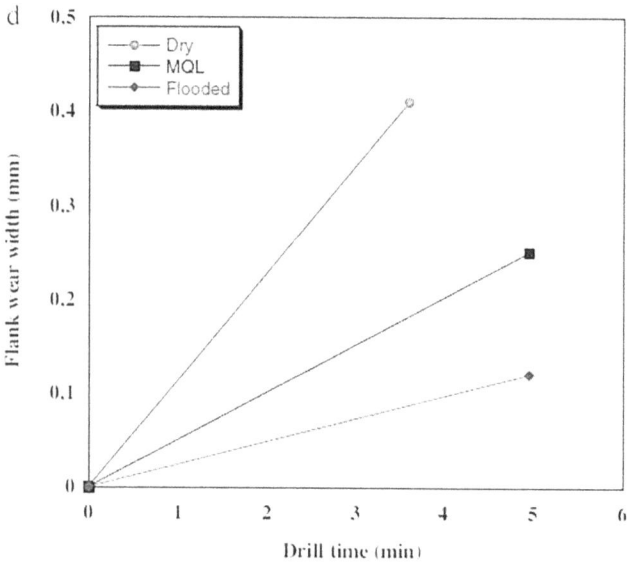

FIGURE 6.28 Comparisons of width of flank wear during ADI drilling under MQL, flooded, and dry conditions (Meena & El Mansori, 2011).

under flooded, MQL, and dry drilling conditions for ADI. From this figure, it is clear that the MQL drilling considerably reduces the tool wear and improves the tool life as compared to dry conditions.

Optical and SEM images of the cutting tool during the drilling of ADI under different cutting conditions are shown in Figure 6.29. From this figure, it was clear that the tool wear was mostly observed near the periphery of the cutting edge. At the beginning of drilling, peeling of tool coating occurs near the cutting-edge periphery, and the WC tool matrix will become directly in contact with the work surface, as displayed in Figure 6.29. As the quantity of drilled holes increases, the frictional force also increases, and built-up-edge (BUE) begins to form. This leads to cutting-edge chipping near the drill periphery. As a result of the crater wear along with the frictional force, BUE, and edge chipping, tool failure occurs. From Figure 6.29, the major tool failure modes are observed as brittle fracture and mechanical breakage during high cutting force with chatter conditions. Similarly, chipping, adhesion, and non-uniform flank wear are observed at the outer cutting edge for both MQL and dry drilling.

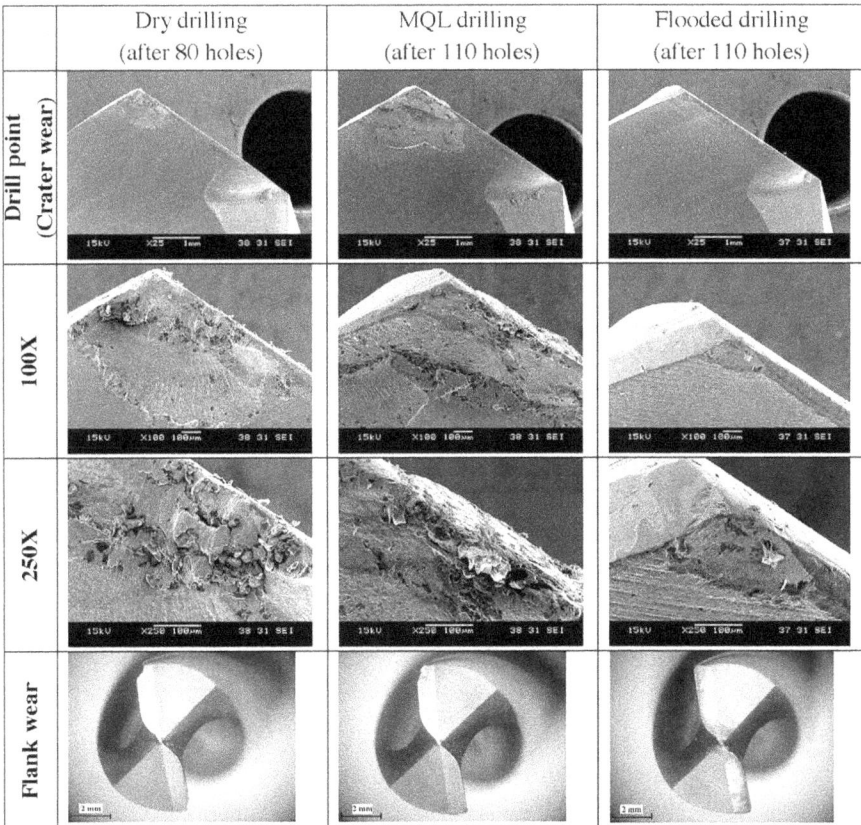

	Dry drilling (after 80 holes)	MQL drilling (after 110 holes)	Flooded drilling (after 110 holes)
Drill point (Crater wear)			
100X			
250X			
Flank wear			

FIGURE 6.29 Images of coated (TiAlN) WC drill bit under MQL, dry, and flooded conditions (Meena & El Mansori, 2011).

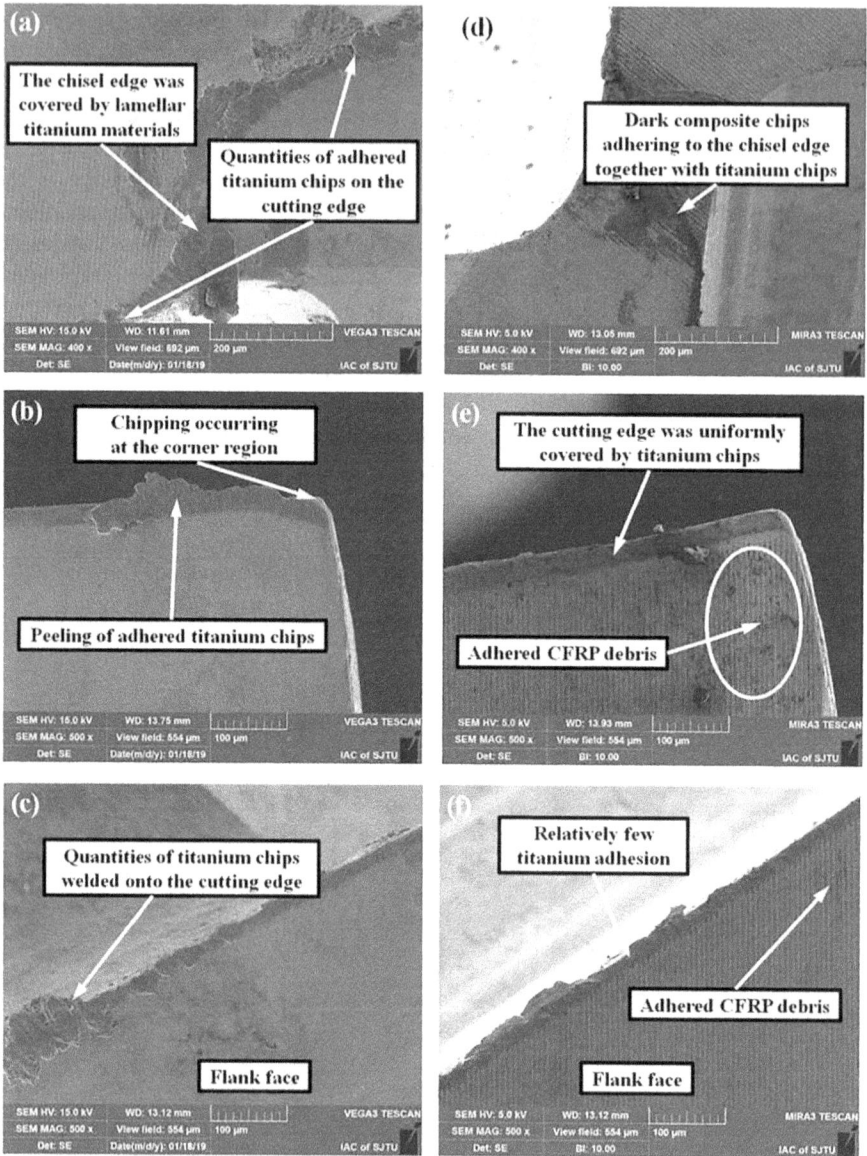

FIGURE 6.30 Tool wear of drill bit for (a-c) dry and (d-f) MQL environments (Xu et al., 2019).

Figure 6.30 shows the wear behavior of the worn tool after drilling 16 holes for dry and MQL conditions during drilling CFRP-Ti6Al4V stacks. Under MQL conditions, both chipping and adhesion are significantly reduced on the cutting edges, as shown in Figure 6.30. In dry drilling, the workpiece chips were highly welded on the cutting edge. Also, severe workpiece adhesion and BUE formation were observed under dry conditions. This results in the delamination of coating at

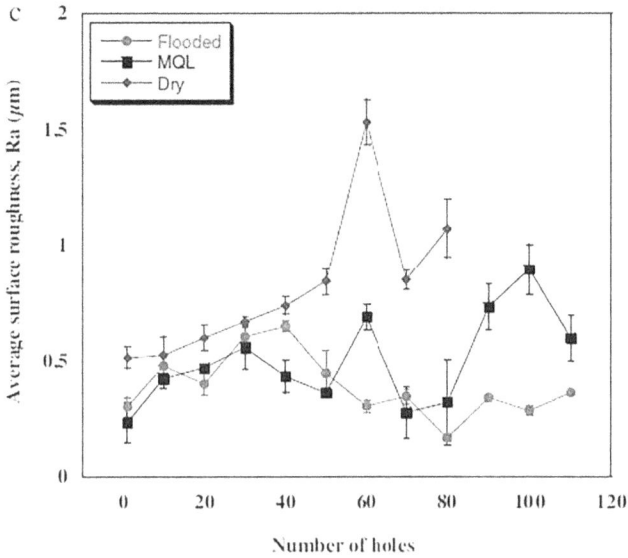

FIGURE 6.31   Evaluations of surface roughness under dry, MQL, and flooded conditions during drilling of ADI (Meena & El Mansori, 2011).

high drilling temperatures. However, for MQL drilling, the powder chips mainly adhere on the tool due to the moisture effect (Xu et al., 2019).

6.4.3   Effect of MQL on surface roughness

In a drilling process, the drilled surface quality has an important role on product acceptability. In this session, the influence of MQL on surface roughness while drilling compared to dry and flood conditions was discussed. Figure 6.31 shows the variation of surface roughness for MQL, flooded, and dry drilling of ADI. From Figure 6.31, it was observed that the surface roughness increases at a very fast rate in dry conditions compared to flood and MQL conditions. This is due to the formation of higher stresses and cutting temperatures near the tool tip. For all cases, the highest value of surface roughness was observed for dry conditions. The roughness value (Ra) for MQL condition is in the range of 0.498 ± 0.2 µm, and the same for flooded conditions is in the range of 0.389 ± 0.136 µm. This indicates that the MQL drilled hole surfaces are a little rougher than that of flooded conditions.

6.4.4   Effect of MQL on chip morphology

The formation of BUE in drilling is very common and has a significant role in surface quality. Due to the adhesion of BUE on the machined surface and chip, the drilled surface quality decreases considerably. Hence, it is very important to avoid the formation of BUE. Continuous chip formation is not desirable in all machining operations. In the case of drilling, the continuous chips may fill the flutes, especially

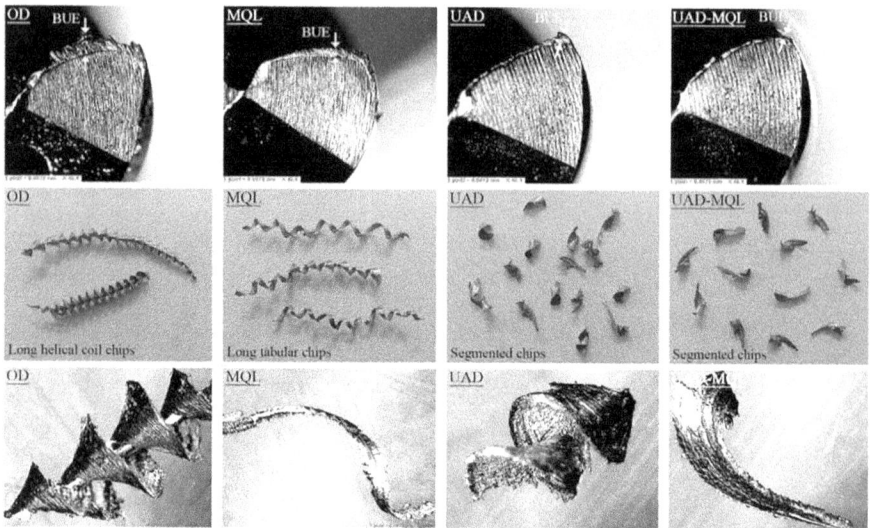

FIGURE 6.32 Chip formation in dry (OD), MQL, ultrasonic-assisted drilling (UAD), and UAD+MQL conditions (Lotfi et al., 2017).

at lower feed ranges and increase the friction, improper cutting, and related issues. This results in an increase in grinding force and a severe reduction in surface quality. Figure 6.32 shows the chip formation during the drilling of AISI 1045 steel. In dry drilling, continuous chips were observed, as shown in Figure 6.32(a). However, in MQL drilling, the chip geometry changes from coil-type helical to more straight chips (Figure 6.32b). But the chips under MQL grinding were not discontinuous due to the lower friction (Lotfi et al., 2015). This issue can be solved by the application of MQL and ultrasonic vibration-assisted drilling (UAD), as shown in Figure 6.32(d). The chips under MQL+UAD were discontinuous. UAD can be used to reduce the Built Up Edge (BUE) formation. As a result, if we require a highly polished surface, MQL+UAD can be used.

6.5 EFFECT OF MQL ON MACHINING PERFORMANCES DURING GRINDING OPERATION

This session discusses the effect of MQL during the grinding operation.

6.5.1 EFFECT OF MQL ON GRINDING FORCE AND SPECIFIC ENERGY

Figure 6.33 displays the variation of grinding force components with flow rates for different air pressure during Inconel 751 machining. It was observed that the increase in lubricant flow decreases the tangential force (Figure 6.33a). However, the normal force shows a different trend, as shown in Figure 6.33(b). The normal force decreases with the flow rate for a lower air pressure value (2 bar). A different trend is observed for 4- and 6-bar air pressure. Around a flow rate of 80 mL/h, a

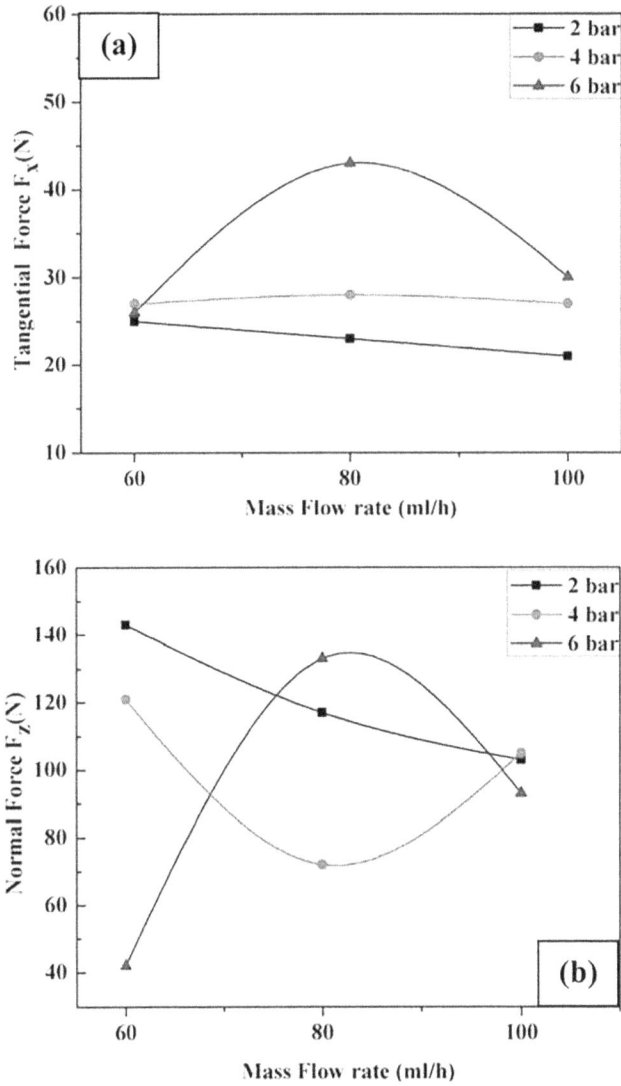

FIGURE 6.33 Effects of (a) tangential force and (b) normal force with flow rate for different air pressure (Balan et al., 2013).

significant change in normal force is observed depending on air pressure. The value of grinding force is highly influenced by specific pressure and temperature (Balan et al., 2013).

6.5.2 EFFECT OF MQL ON GRINDING TEMPERATURE

Figure 6.34 shows the variation in grinding temperature for different MQL flow rates. From this figure, a different trend for grinding temperature with flow rate

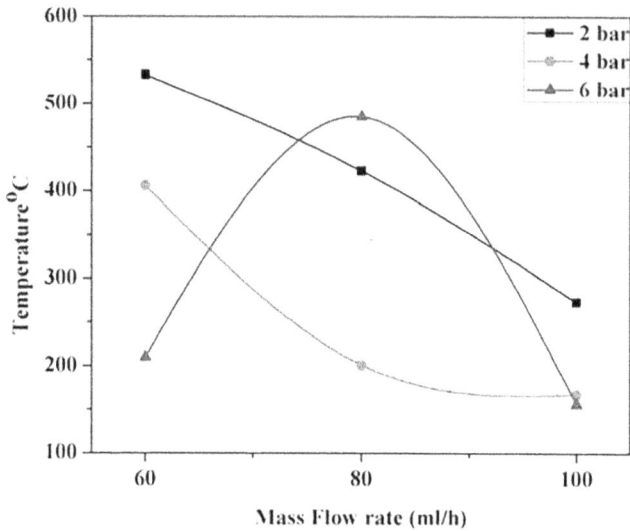

FIGURE 6.34 Effects of grinding temperature on mass flow rate for different air pressure (Balan et al., 2013).

under different air pressure was observed. For both 2-bar and 4-bar air pressure, the grinding temperature decreases with an increased flow rate. However, at 6-bar air pressure, the grinding temperature increases up to 80 mL/h flow rate and then decreases. The reduction in grinding temperature for 2-bar pressure leads to a decrease in grinding force, as shown in Figure 6.34. But for 4-bar pressure, to 80 mL/h flow rate, the grinding temperature decreases and then increases. This change is reflected in normal force. Similarly, for 6-bar pressure, the force components change (Figure 6.34) in alignment with the grinding temperature (Figure 6.34). The maximum grinding temperature observed at 6-bar pressure is observed at 80 mL/h flow rate. This may be due to the variation in MQL velocity and oil droplet size. Hence, the efficiency of the MQL system highly depends on these factors. That means these factors need to be carefully chosen to get the optimum machining performance. In the grinding process, a belt air pressure is created. The proper wetting of the grinding wheel depends on the lubrication ability of the MQL to penetrate, depending upon the droplet size and velocity. This penetration efficiency to obtain minimum MQL-assisted grinding temperature increases with a decrease in size and increase in velocity of the droplet.

As the flow rate and air pressure increase, the size of the droplet reduces with an increase in velocity (Balan et al., 2013). Hence, the higher air pressure and flow rate lead to the effective penetration lubricant and improves the wetting behavior of the interface. As a result, the grinding temperature decreases with pressure and flow rate. But, for air pressure up to 4 bar and a flow rate of 80 mL/h, a variation in trend is observed. This is due to the partial atomization of cutting fluid with air pressure. Hence, the fragmentation of droplets occurs and results in a decrease in accessibility and wetting of the cutting zone. Hence, the grinding temperature increases above

80 mL/h for 6-bar pressure. The increase in grinding temperature softens the bond and results to increase in normal force. Also, at a flow rate of 100 mL/h, only a small difference in grinding temperature is observed for both 4- and 6-bar pressure. The same observation is reflected in forces.

6.5.3 EFFECT OF MQL ON SURFACE ROUGHNESS

Figure 6.35 shows the effect of the MQL flow rate on surface roughness (Ra). It was observed that the increase in flow rate decreases the surface roughness for pressure up to 4-bar, and at 6-bar pressure, the roughness increases with the MQL flow rate. For 60 mL/h flow rate, surface roughness decreases with an increase in air pressure. But only a small variation in surface roughness was observed for 100 mL/h flow rate. For both 2- and 4-bar pressure, surface roughness decreases to 80 mL/h and then increases. However, for l-bar pressure, surface roughness increases with a flow rate to 80 mL/h and at 100 mL/h, surface roughness becomes steady. It was also found that the influence of air pressure is minimum at a high flow rate (100 mL/h). But at a low flow rate (60 mL/h), the surface roughness is extremely sensitive to air pressure as shown in Figure 6.35. For 80 mL/h, both high pressure (6 bar) and low pressure (2 bar) show a high roughness value. At a low flow rate, the grinding temperature will be high and which leads to an increase in roughness value. Similarly, as the flow rate increases, smooth grinding action occurs. 3D profile of the Inconel 751 surface for high and low Ra values was shown in Figure 6.35. In general, at higher pressure, a small droplet mist is formed and results in the efficient penetration of cutting fluid.

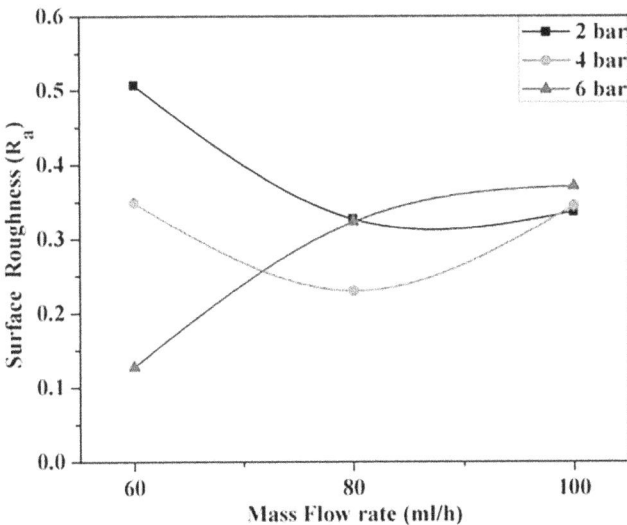

FIGURE 6.35 Effects of surface roughness on mass flow rate for different air pressure (Balan et al., 2013).

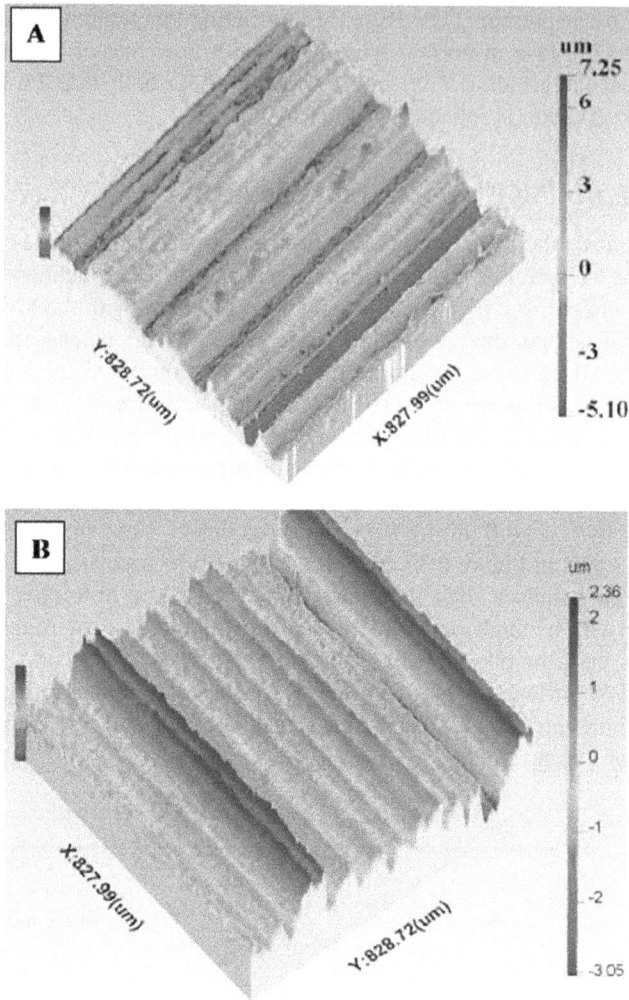

FIGURE 6.36   Machined Inconel surface for (a) high and (b) low Ra values (Balan et al., 2013).

The machined surface morphology of Inconel with the application of MQL for high and low roughness values is shown in Figure 6.36. In Figure 6.37(a), a significant plastic flow on the machined surface due to high grinding temperature is observed. In this case, lubrication is observed to be not effective and the surfaces are very uneven with surface defects. Figure 6.37(b) shows a smooth surface obtained due to the effective lubrication and easy removal of the adhered chip on the grinding wheel or workpiece. Hence, a better surface quality is observed as shown in Figure 6.37(b). The application of MQL at high air pressure is can penetrate easily to the cutting zone. Also, a reduction in surface energy due to the effective MQL leads to an increase in elastic-plastic deformation (El Shall & Somasundaran, 1984).

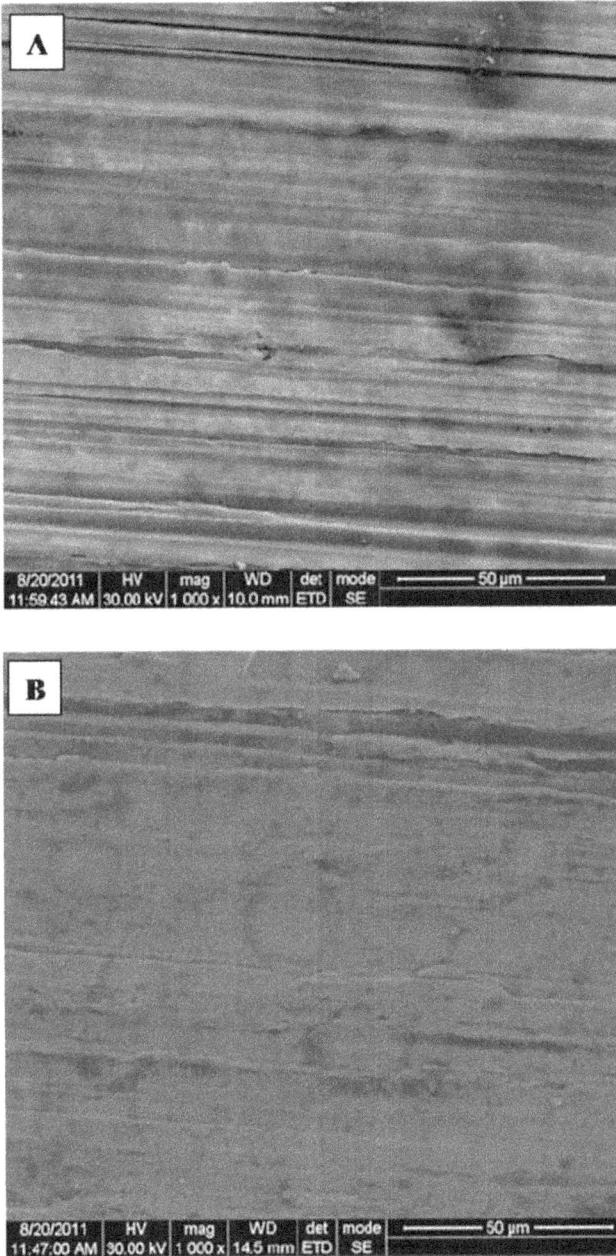

FIGURE 6.37   Inconel surface for (a) high and (b) low Ra value (Balan et al., 2013).

6.5.4   EFFECT OF MQL ON MICROHARDNESS

As far as surface integrity and application point of view, the analysis of machined surface microhardness is very important. Awale et al. (2019) investigated the effect

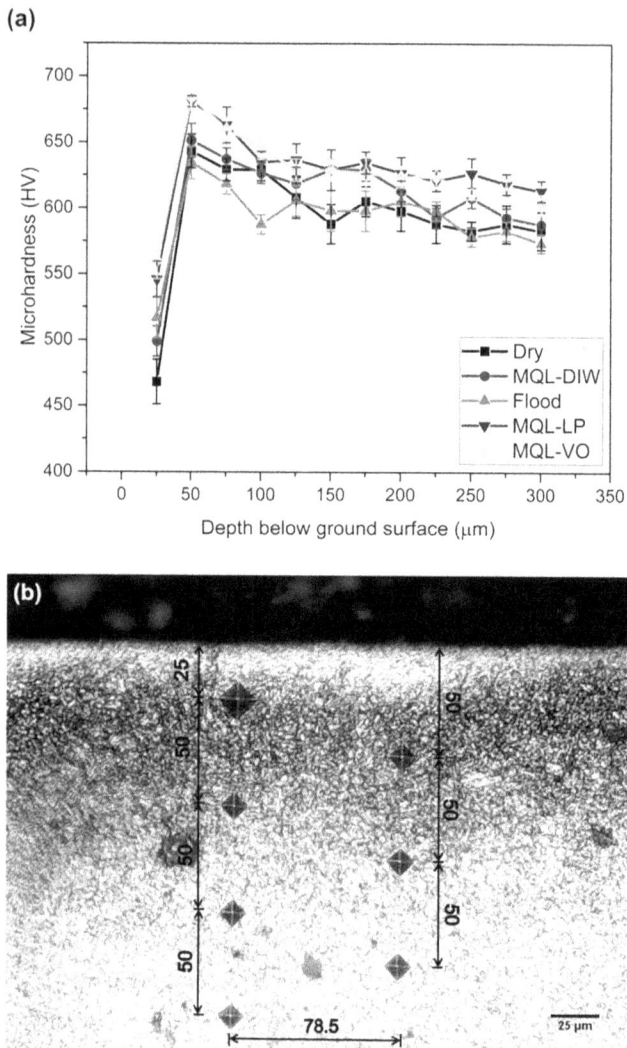

FIGURE 6.38 (a) Variation of microhardness with distance from the machined surface for dry, flood and MQL with deionized water (DIW), liquid paraffin (LP) and vegetable oil (VO) (b) location of microhardness measuring points (Awale et al., 2019).

of MQL on the microhardness of machined surfaces during the grinding of die steel for five different cutting conditions. The understanding of microhardness can be used to categorize the phases in the material. Figure 6.38(a) shows the variation of microhardness with distance from the machined surface and Figure 6.38(b) shows the location of indentations or hardness measurement points. In dry grinding, a low value of microhardness is observed compared to other conditions due to the fast heating and slow cooling of the machined surfaces. In a grinding process, the amount of carbon diffusion depends on the grinding time and temperature (Silva et al., 2020). The high feed and low wheel velocity lead to the development of huge heat, and this will result to the reduction in microhardness (Shaw & Vyas, 1994). For deionized water-based MQL, a similar kind of trend is observed due to insufficient lubrication. For flooded grinding, a small variation in microhardness is observed due to the sufficient lubrication effect. MQL using vegetable oil (VO) system, shows a high hardness value compared to other conditions due to the better friction, wear, and lubrication properties. As a result, better cooling and lubrication efficiency was obtained for elastic as well as plastic conditions. The reduction in grinding temperature leads to an increase in microhardness. Huang et al. (Huang et al., 2017) also found that the MQL grinding gives a high surface microhardness with flood and dry grinding.

6.5.5 Effect of MQL on grinding wheel wear

Grinding wheel condition is an essential parameter that significantly affects the quality of the machined components. This session deals with the investigations related to the effect of MQL on grinding wheel wear. In general, bond fracture and abrasive grain fracture are significant factors that influence grinding wheel wear (Tawakoli et al., 2011). Higher cutting temperatures reduce the bond strength and leads to abrasive grain removal and bond fracture. Also, the wheel porosity increases power and stress, and decreases durability (Bianchi et al., 2018). The high grinding temperature may reduce the mechanical resistance of abrasive grits and decreases the material removal rate. The abrasive particles loses its mechanical resistance as a result of the increase in cutting temperature (Tawakoli et al., 2011). Figure 6.39 shows the effect of different lubrication strategies on diametric wheel wear during the grinding of hardened steel using a CBN wheel.

From Figure 6.39, it is clear that compared to flood cooling, the application MQL causes high wheel wear for all conditions. This is due to the lower cooling ability and high clogging of MQL-assisted grinding (de Oliveira et al., 2012). As a result,

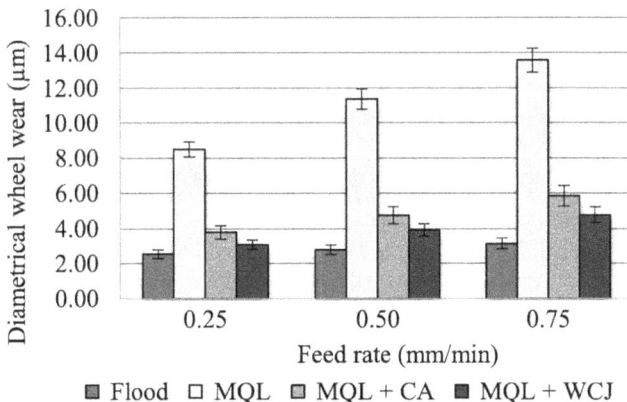

FIGURE 6.39   Variation of diametrical wheel wear for flood, MQL, MQL + cold air at 0°C (CA) and MQL + wheel cleaning jet (WCJ) different lubrication strategies during grinding of hardened steel using a CBN wheel (Lopes, et al., 2019).

abrasive grain resistance and bonding decrease, and the grinding temperature between the wheel and workpiece increases. This increases the cutting energy and increases the wheel wear significantly. Due to the superior cooling, lubrication, and chip removal ability of the flood lubrication system, the wheel wear will be very low compared to other systems. The flood cooling reduces both cutting temperature and wheel clogging (Tawakoli et al., 2011). The MQL+CA (cold air) condition shows a substantial diametrical wheel wear reduction due to the softening of the grinding wheel (Tawakoli et al., 2011). The MQL+WCJ shows the lowest diametrical wheel wear in comparison with all other systems. Hence, it can be noted that along with grinding temperature reduction, chip removal is also very important while discussing wheel wear. Hence, both MQL+CA and MQL+WCJ can be used as an alternative to flood system and saves a huge amount of lubricant.

REFERENCES

Anand, K. N., & Mathew, J. (2020). Evaluation of size effect and improvement in surface characteristics using sunflower oil-based MQL for sustainable micro-endmilling of Inconel 718. *J. Brazilian Society* of *Mechanical Sciences* and *Engineering*, 42(4), 1–13.

Anand, K. N., & Mathew, J. (2020). Studies on wear behavior of AlTiN-coated WC tool and machined surface quality in micro endmilling of Inconel 718. *International Journal* of *Advanced Manufacturing Technology*, 110, 291–307.

Anand, K. N., & Mathew, J. (2021). Size effect and micro endmilling performance while sustainable machining on Inconel 718. *Materials and Manufacturing Processes*, 36(6), 668–676.

Awale, A. S., Srivastava, A., Vashista, M., & Yusufzai, M. Z. K. (2019). Influence of minimum quantity lubrication on surface integrity of ground hardened H13 hot die steel. *International Journal* of *Advanced Manufacturing Technology*, 100, 983–997.

Balan, A. S. S., Vijayaraghavan, L., & Krishnamurthy, R. (2013). Minimum quantity lubricated grinding of Inconel 751 Alloy. *Materials and Manufacturing Processes*, 28(4), 430–435.

Barczak, L. M., Batako, A. D. L., & Morgan, M. N. (2010). A study of plane surface grinding under minimum quantity lubrication (MQL) conditions. *International Journal of Machine Tools and Manufacture*, 50, 977–985.

Behera, B. C., Ghosh, S., & Rao, P. V. (2018). Modeling of cutting force in MQL machining environment considering chip tool contact friction. *Tribology International*, 117, 283–295.

Benjamin, D. M., Sabarish, V. N., Hariharan, M. V., & Samuel R. D. (2018). On the benefits of sub-zero air supplemented minimum quantity lubrication systems: An experimental and mechanistic investigation on end milling of Ti6Al4V Alloy. *Tribology International*, 119, 464–473.

Bianchi E. C., Sato, B. K., Sales, A. R., Lopes, J. C., de Mello, H. J., Sanchez, A. L. E., Diniz, A. E., & Aguiar, P. R. (2018). Evaluating the effect of the compressed air wheel cleaning in grinding the AISI 4340 steel with CBN and MQL with water. *International Journal* of *Advanced Manufacturing Technology*, 95(5–8), 2855–2864.

Bordin, A., Bruschi, S., & Ghiotti, A. (2014). The effect of cutting speed and feed rate on the surface integrity in dry turning of CoCrMo alloy. *Procedia CIRP*, 13, 219–224.

Boswell, B., Islam, M., Davies, I. J., Ginting, Y. R., & Ong, A. K. (2017). A review identifying the effectiveness of minimum quantity lubrication (MQL) during conventional machining. *International Journal* of *Advanced Manufacturing Technology*, 92, 321–340.

Brinkmeier, E., Brockoff, T., & Walter, A. (1997). Minimum quantity lubrication in grinding. Society of *Manufacturing Engineers*, 97(230), 1–14.

Cakir, A., Yagmur, S., Kavak, N., Kucukturk, G., & Seker, U. (2016). The effect of minimum quantity lubrication under different parameters in the turning of AA7075 and AA2024 aluminium alloys. *International Journal* of *Advanced Manufacturing Technology*, 84(9–12), 2515–2521.

Ciftci, I. (2006). Machining of austenitic stainless steels using CVD multi-layer coated cemented carbide tools. *Tribology International*, 39, 565–569.

Da Silva, L. R., Bianchi, E. C., Fusse, R. Y., Catai, R. E., França, T. V., & Aguiar, P. R. (2007). Analysis of surface integrity for minimum quantity lubricant-MQL in grinding. *International Journal of Machine Tools and Manufacture*, 47 (2), 412–418.

Das, S. R., Kumar, A., & Dhupal, D., (2016). Experimental investigation on cutting force and surface roughness in machining of hardened AISI 52100 steel using CBN tool. *International Journal of Machining and Machinability of Materials*, 18(5/6), 501–521.

Davim, J. P., Sreejith, P. S., & Silva, J. (2007). Turning of brasses using minimum quantity of lubricant (MQL) and flooded lubricant conditions. *Materials and Manufacturing Processes*, 22(1), 45–50.

de Oliveira, D. J., Guermandi, L. G., Bianchi, E. C., Diniz, A. E., de Aguiar, P. R., & Canarim, R. C. (2012). Improving minimum quantity lubrication in CBN grinding using compressed air wheel cleaning. *Journal of Materials Processing Technology*, 212(12), 2559–2568.

Derani, M. N., & Ratnam, M. M. (2021). The use of tool flank wear and average roughness in assessing effectiveness of vegetable oils as cutting fluids during turning—A critical review. *International Journal* of *Advanced Manufacturing Technology*, 112, 1841–1871.

Dhar, N. R., Ahmed, M. T., & Islam, S. (2007). An experimental investigation on effect of minimum quantity lubrication in machining AISI 1040 steel. *International Journal of Machine Tools and Manufacture*, 47(5), 748–753.

Dhar, N. R., Islam, M. W., Islam, S., & Mithu, M. A. H. (2006). The influence of minimum quantity of lubrication (MQL) on cutting temperature, chip and dimensional accuracy in turning AISI-1040 Steel. *Journal of Materials Processing Technology*, 171(1), 93–99.

Dhar, N. R., Kamruzzaman, M., & Ahmed, M. (2006). Effect of minimum quantity lubrication (MQL) on tool wear and surface roughness in turning AISI-4340 steel. *Journal of Materials Processing Technology*, 172(2), 299–304.

Diniz, A. E., Ferreira, J. R., & Filho, F. T. (2003). Influence of refrigeration/lubrication condition on SAE 52100 hardened steel turning at several cutting speeds. *International Journal of Machine Tools and Manufacture*, 43(3), 317–326.

Do, T. V., & Hsu, Q. C. (2016). Optimisation of minimum quantity lubricant conditions and cutting parameters in hard milling of AISI H13 steel. *Applied Sciences*, 6, 1–11.

El Shall, H., & Somasundaran, P. (1984). Physico-chemical aspects of grinding: A review of use of additives. *International Journal of Powder Technology*, 38, 275–293.

Gurbuz, H., Gonulacar. Y. E., & Baday, S. (2020). Effect of MQL flow rate on machinability of AISI 4140 steel. *Machining Science and Technology*, 24(5), 663–687.

Hassanpour, H., Rasti, A., Sadeghi, M. H., & Khosrowshahi, J. H. (2020). Investigation of roughness, topography, microhardness, white layer and surface chemical composition in high speed milling of Ti-6Al-4V using minimum quantity lubrication. *Machining Science and Technology*, 24(5), 719–738.

Hou, J., Zhou, W., Duan, H., Yang, G., Xu, H., & Zhao, N. (2014). Influence of cutting speed on cutting force, flank temperature, and tool wear in end milling of Ti-6Al-4V alloy. *International Journal* of *Advanced Manufacturing Technology*, 70(9–12), 1835–1845.

Huang, X. M., Ren, Y. H., Jiang, W., He, Z. J., & Deng, Z. H. (2017). Investigation on grind-hardening annealed AISI5140 steel with minimal quantity lubrication. *International Journal* of *Advanced Manufacturing Technology* 2017, 89(1-4), 1069–1077.

Hwang, Y. K., & Lee, C. M., (2010). Surface roughness and cutting force prediction in MQL and wet turning process of AISI 1045 using design of experiments. *Journal of Mechanical Science and Technology*, 24(8), 1669–1677.

Jayal, A. D., Badurdeen, F., Dillon, O. W., & Jawahir, I. S., (2010). Sustainable manufacturing: Modeling and optimisation challenges at the product, process and system levels. *CIRP Journal Manuf. Sc. Tech.*, 2(3), 144–152.

Ji, X., Li, B., & Zhang, X. (2014). The effects of minimum quantity lubrication (MQL) on machining force, temperature, and residual stress. *International Journal of Precision Engineering and Manufacturing*, 15, 2443–2451.

Ji, X., Li, B., Zhang, X., & Liang, S. Y. (2014). The effects of minimum quantity lubrication (MQL) on machining force, temperature, and residual stress. *International Journal of Precision Engineering and Manufacturing*, 15(11), 2443–2451.

Kalsi, N. S., Sehgal, R., & Sharma. V. S. (2010). Cryogenic treatment of tool materials: A review. *Materials and Manufacturing Processes*, 25, 1077–1100.

Khan, A., & Maity, K., (2018). Influence of cutting speed and cooling method on the machinability of commercially pure titanium (CP-Ti) grade II. *Journal of Manufacturing Processes*, 31, 650–661.

Lopes, J. C., Fragoso, K. M., Garcia, M. V., Ribeiro, F. S. F., Francelin, A. P., Sanchez, A. L. E., Rodrigues, A. R., de Mello, H. J., Aguiar, P. R., & Bianchi, E. C. (2019). Behavior of hardened steel grinding using MQL under cold air and MQL CBN wheel cleaning. *International Journal of Advanced Manufacturing Technology* 2019, 105(10), 4373–4387.

Lotfi, M., Amini, S., Teimouri, R., & Alinaghian, M. (2017). Built-up edge reduction in drilling of AISI 1045 Steel. *Materials and Manufacturing Processes*, 32(6), 623–630.

Lotfi, M., Farid, A. A., & Soleimanimehr, H. (2015). The effect of chip breaker geometry on chip shape, bending moment, and cutting force: FE analysis and experimental study. *International Journal of Advanced Manufacturing Technology*, 78(5-8), 917–925.

Mallesha, V., & Nayaka, H. S. (2018). Turning process on EN47 spring steel with different tool nose radii using OFAT approach. *Advances in Modelling Analysis*, 55(2), 43–46.

Masoudi, S., Vafadar, A., Hadad, M., & Jafarian, F. (2018). Experimental investigation into the effects of nozzle position, workpiece hardness, and tool type in MQL turning of AISI 1045 steel. *Materials and Manufacturing Processes*, 33(9), 1011–1019.

Meena, A., & El Mansori, M. (2011). Study of dry and minimum quantity lubrication drilling of novel austempered ductile iron (ADI) for automotive applications. *Wear*, 271(9-10), 2412–2416.

Morgan, M. N., Barczak, L., & Batako, A. (2012). Temperatures in fine grinding with minimum quantity lubrication (MQL). *International Journal of Advanced Manufacturing Technology*, 60, 951–958.

Nithin, T. M., & Vijayaraghavan, L. (2018). Temperature rise in workpiece and cutting tool during drilling of titanium aluminide under sustainable environment. *Materials and Manufacturing Processes*, 33(16), 1765–1774.

Narayanan, S. V., Benjamin, D. M., Hariharan, M. V., Keshav, R., & D. Samuel, R. (2021). A combined numerical and experimental investigation of minimum quantity lubrication applied to end milling of Ti6Al4V alloy. *Machining Science and Technology*, 25, 209–236.

Obikawa, T., Kamata, Y., Asano, Y., Nakayama, K., & Otieno, A. W. (2008). Micro-liter lubrication machining of Inconel 718. *International Journal of Machine Tools and Manufacture*, 48, 1605-161.

Pathak, B. N., Sahoo, K. L., & Mishra, M. (2013). Effect of machining parameters on cutting forces and surface roughness in Al-(1-2) Fe-1V-1Si Alloys. *Materials and Manufacturing Processes*, 28(4), 463–469.

Patole, P. B., & Kulkarni, V. V. (2017). Experimental investigation and optimisation of cutting parameters with multi response characteristics in MQL turning of AISI 4340 using nano fluid. *Cogent Eng.* 2017, 4(1), 1303956.

Pervaiz, S., Anwar, S., Qureshi, I., & Ahmed, N. (2019). Recent advances in the machining of titanium alloys using minimum quantity lubrication (MQL) based techniques. *International Journal of Precision Engineering and Manufacturing Green Technology*, 6, 133–145.

Pramanik, A., & Littlefair, G. (2015). Machining of titanium alloy (Ti-6Al-4V)-theory to application. *Machining Science and Technology*, 19(1), 1–49.

Rahman, M., Senthil, K. A., Salam, M. U. (2002). Experimental Evaluation on the Effect of Minimal Quantities of Lubricant in Milling. *International Journal of Machine Tools and Manufacture*, 42, 539–547.

Rajaguru, J., & Arunachalam, N. (2020). A comprehensive investigation on the effect of flood and MQL coolant on the machinability and stress corrosion cracking of super duplex stainless steel. *J Mat. Process. Tech.*, 276, 116417.

Revuru, R. S., & Rao, P. N. (2017). Use of vegetable-based cutting fluids for sustainable machining. In: Davim, J. (Eds), *Sustainable Machining. Materials Forming, Machining and Tribology* (pp. 31–46). Cham: Springer, .

Revuru, R. S., Zhang, J. Z., & Rao, P. N. (2020). Comparative performance studies of turning 4140 Steel with TiC/TiCN/TiN-coated carbide inserts using MQL, flooding with vegetable cutting fluids, and dry machining. *International Journal of Advanced Manufacturing Technology*, 108(1–2), 381–391.

Salur, E., Kuntoğlu, M., Aslan, A., & Pimenov, D. Y. (2021). The effects of MQL and dry environments on tool wear, cutting temperature, and power consumption during end milling of AISI 1040 steel. *Metals*, 11(11), 1674.

Sani, A. S. A., Rahim, E. A., Sharif, S., & Sasahara, H. (2019). The influence of modified vegetable oils on tool failure mode and wear mechanisms when turning AISI 1045. *Tribology International*, 129, 347–362.

Sani, A. S. A., Rahim, E. A., Sharif, S., & Sasahara, H. (2019). Machining performance of vegetable oil with phosphonium and ammonium based liquids via MQL technique. *Journal of Cleaner Production*, 209, 947–964.

Sarikaya, M., & Gullu, A., (2014). Taguchi design and response surface methodology based analysis of machining parameters in CNC turning under MQL. *Journal of Cleaner Production*, 65, 604–616.

Sarikaya, M., & Gullu, A. (2015). Multi-response optimization of minimum quantity lubrication parameters using taguchi-based grey relational analysis in turning of difficult-to-cut alloy Haynes 25. *Journal of Cleaner Production*, 91, 347–357.

Schoop, J., Sales, W. F., & Jawahir, I. S. (2017). High speed cryogenic finish machining of Ti-6Al4V with polycrystalline diamond tools. *Journal of Materials Processing Technology*, 250, 1–8.

Sharma, V. S., Dogra, M., & Suri, N. M. (2009). Cooling techniques for improved productivity in turning. *International Journal of Machine Tools and Manufacture*, 49, 435–453.

Shaw, M., & Vyas, A. (1994). Heat-affected zones in grinding steel. *CIRP Annals*, 43(1), 279–282.

Silva, L. R., Corrêa, E. C., Brandão, J. R., & de Ávila, R. F. (2020). Environmentally friendly manufacturing: behavior analysis of minimum quantity of lubricant-MQL in grinding process. *Journal of Cleaner Production*, 256, 103287.

Sun, J., Wong, Y. S., Rahman, M., Wang, Z. G., Neo, K. S., Tan, C. H., & Onozuka, H. (2006). Effects of coolant supply methods and cutting conditions on tool life in end milling titanium alloy. *Machining Science and Technology*, 10(3), 355–370.

Sun, Y., Huang, B., Puleo, D. A., & Jawahir, I. S. (2015). Enhanced machinability of Ti-5553 alloy from cryogenic machining: Comparison with MQL and flood-cooled machining and modeling. *Procedia CIRP*, 31, 477–482.

Tai, B. L., Stephenson, D. A., Furness, R. J., & Shih, A. J. (2014). Minimum quantity lubrication (MQL) in automotive powertrain machining. *Procedia CIRP*, 14, 523–528.

Tawakoli, T., Hadad, M., Sadeghi, M. H., Daneshi, A., & Sadeghi, B. (2011). Minimum quantity lubrication in grinding: effects of abrasive and coolant–lubricant types. *Journal of Cleaner Production*, 19(17–18), 2088–2099.

Vishal, S. S., Singh, G., & Sørby, K. (2015). A Review on Minimum Quantity Lubrication for Machining Processes. *Materials and Manufacturing Processes*, 30(8), 935–953.

Weck, M., & Koch, A. (1993). Spindle bearing systems for high-speed applications in machine tools. *CIRP Annals - Manufacturing Technology*, 42(1), 445–448.

Xu, J., Ji, M., Chen, M., & Ren, F. (2019). Investigation of minimum quantity lubrication effects in drilling CFRP/Ti6Al4V stacks. *Materials and Manufacturing Processes*, 34(12), 1401–1410.

7 Applications of MQL in micro-machining

7.1 INTRODUCTION

Minimum quantity lubrication (MQL) refers to the supply of compressed air mixture with a minimal quantity of cutting oil usually at a flow rate of 50–500 mL/h (Dhar et al., 2006). The amount of lubrication supplied is usually three to four times lower than the magnitude of fluid supplied in flood cooling conditions. While the use of the MQL technique not only reduces the machining cost, it also facilitates in reducing the issues related to environmental hazards in terms of its disposal and health-related issues in personnel operating the machine. As the cutting oil is supplied in a minimum desirable proportionate mixture with the compressed air, the issues experienced during machining with uncontrolled usage of lubricants as in the case of flood cooling and high-pressure jet-assisted flood cooling, can be minimized to a large extent. While the compressed air effectively acts as a coolant, the presence of the minimum quantity of cutting oil mixed with the air helps in facilitating the role of an effective lubricant that assists in minimizing friction and the extreme temperatures at the contact interface during machining. The major issue faced during machining under flood cooling condition is that although the cutting fluid may be able to penetrate the tool chip contact interface during machining at lower cutting speeds, its penetration at medium to higher cutting speeds are severely limited in the case of medium to higher cutting speeds, especially in the case of hard-to-machine elements like superalloys. The reason may be attributed to the fact that during the machining of superalloys with a reduced degree of thermal conductivity at higher cutting speeds, the conventional cutting fluids may not be able to break into the tool work contact interface gap due to the possible effect of the Leidenfrost phenomenon. As the cutting fluid comes in contact with the surface temperature severely higher than its boiling point, an insulating blanket-like layer is formed which limits its penetration across the machining interface. However, the presence of minimum quantity lubrication in mist air form will mostly be able to penetrate across the interface and vaporize after absorbing a sufficient quantity of heat from the machining interface (Mahesh et al., 2021). Although the MQL technique offers numerous advantages as mentioned above in terms of minimal usage of cutting fluids and addressing issues related to environmental hazards, the cooling features may not be up to the required levels (Patole & Kulkarni, 2018). In such a scenario, the incorporation of solid lubricants like Al_2O_3, MWCNT, MoS_2, etc. with the compressed air-oil mixture may be implemented to improve the cooling and heat dissipation efficiency of the fluids during the machining. The nanosized particles commonly possess a higher thermal conductivity and exhibit superior heat dissipation characteristics. Moreover, these particles when used in

DOI: 10.1201/9781003328742-7

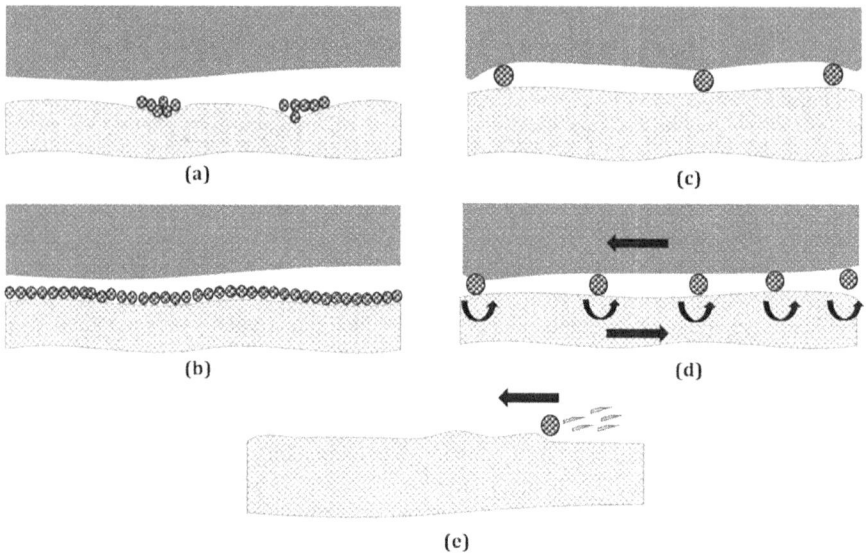

FIGURE 7.1 Possible tribological enhancing mechanisms of nanoparticle-assisted MQL technique (Öndin et al., 2020).

combination with MQL display a higher load-bearing capacity at the machining interface, which helps in facilitating reduced frictional characteristics during machining in addition to providing a polishing action along the surface of the workpiece which imparts an improved surface finish to the machined product (Öndin et al., 2020). Figure 7.1 displays the beneficial tribological enhancing mechanisms of nano-particle assisted MQL techniques. Figure 7.1a describes the condition in which the nanosized particles settle into the pores of the workpiece and repair the cracked surface by filling them, thereby reducing the contact pressure at the interface. Figure 7.1a and Figure 7.1b describe the tendency of particles to form a protective layer of film between the surfaces and its general load-bearing capacity, respectively. While Figure 7.1c illustrates the rolling tendency of the nanosized particles at the interface between the tool and the chip surface that eventually results in reduced friction constraints, Figure 7.1d details about the polishing tendency of the particle whereby it removes residues present on the surface of the particle which contributes to an effective improvement in surface finish across the machined specimen.

7.2 APPLICATIONS OF MQL IN MICRO-MACHINING

7.2.1 Micro-milling

Recently, numerous works on micro-machining studies of difficult-to-cut materials like nickel- and titanium-based superalloys have been conducted on account of their enormous applications in vast areas of biomedical, aerospace, optics, and electronics applications (Masuzawa, 2000). Machining these components is extremely difficult, in spite of its advantages in terms of its excellent dynamic shear strength, superior corrosion, and oxidation resistance, high hot hardness, as well as strength even at elevated strengths. However, low thermal conductivity, high chemical reactivity and severe work hardening characteristics result in its poor machinability characteristics (Manikandan et al., 2019). In such a scenario, the application of MQL performs a vital role in improving the machinability characteristics of the superalloy. Anand et al. studied the phenomenon of size effect and examined the surface improvement characteristics during micro-end milling of Inconel 718 (Anand & Mathew, 2020). The authors explored the usage of sunflower-based oil as a minimum quantity lubricant while machining the super alloy due to its biode-gradability and its improved tribological properties in terms of its effective cooling and reduced frictional contact at the machining interface. Figure 7.2 shows the enlarged view of the MQL setup in the micro-machining center. Figure 7.3a and Figure 7.3b represent graphs showing the comparison of aerial surface roughness and top burr heights at various feed rates at dry and MQL conditions. They

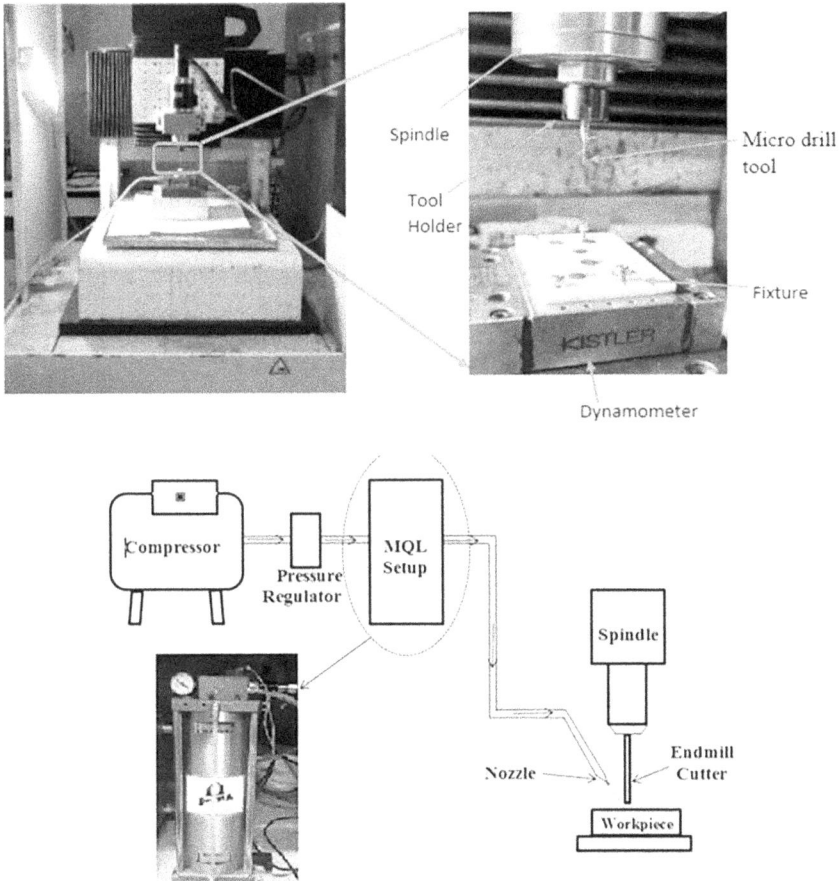

FIGURE 7.2 View of the micro-machining center with a micro-drilling setup (Joshy & Kuriachen, 2023) and MQL setup (Anand & Mathew, 2020).

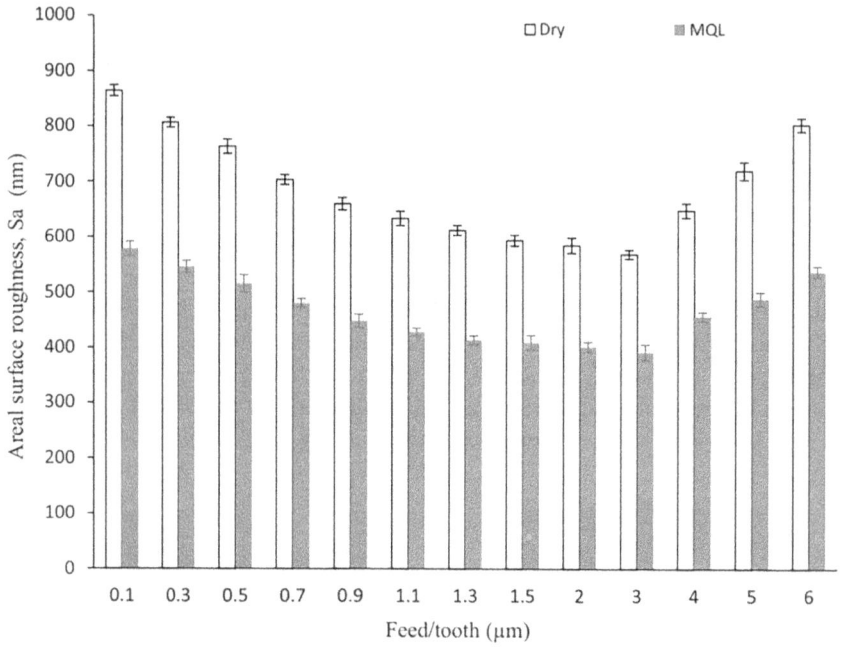

(a)

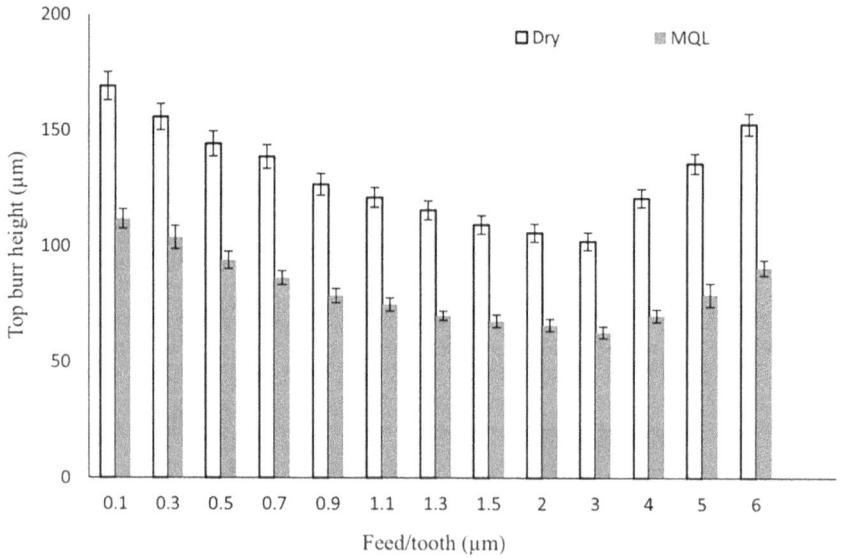

(b)

FIGURE 7.3 Graphs shows the comparison between dry and MQL conditions for (a) areal surface roughness and (b) top burr height for various feed rates (Anand & Mathew, 2020).

observed an overall average reduction in aerial surface roughness by 32%, an average reduction in top burr width and burr height by 36% and 37%, respectively, in machining with the assistance of MQL conditions in contrast to machining the alloy under dry milling. Hassanpour et al. analyzed the micro-machinability characteristics of Ti-6Al-4V in MQL conditions. The authors used mineral oil as the minimum quantity lubricant by supplying the coolant at a flow rate of 240 mL/h and at an air pressure of 6 bars. Significant improvement in machined surface quality was observed under MQL-assisted lubricating conditions (Hassanpour et al., 2016). Anand et al. conducted micro-milling experiments on Inconel 718 with sunflower oil as the minimum quantity lubricant and compared its performance with that machining under dry lubrication conditions. An effective reduction in edge radius, tool flank, and cutting forces by 25%, 30%, and 36%, respectively, in MQL-assisted conditions in contrast to dry machining conditions. Moreover, an increase in microhardness by around 7% with reduced surface defects like cracks, prows, and voids was observed while machining with MQL condition (Anand & Mathew, 2021).

Ucun et al. analyzed the performance of the AlCrN coated tool in micro-milling of Inconel 718 at dry, cryogenic cooling, and MQL cooling conditions. A vegetable-based liquid lubricant was used as the minimum quantity lubricant. Tool wear, the surface roughness of the specimen, and surface burr generation were measured during the machining of the alloy under various lubrication conditions (Ucun et al., 2015).

Based on the results obtained, it was inferred that minimum tool wear was observed in the MQL condition, followed by cryogenic cooling and dry machining, respectively. The reason may be attributed to the fact that in cryogenic cooling, the embrittlement of the workpiece surface can be obtained, which can result in an accelerated rate of tool wear due to increased contact pressure at the machining surface as compared to MQL condition.

Kajaria et al. performed micro-milling experiments on 316L stainless steel in dry and MQL machining conditions. The study explored the usage of commonly available oil-based fluids like CL2200, CL2210, CL2210EP, and CL2300HD. They observed that the cumulative life tests displayed an increase in tool life characteristics by 100 times when machined using CL2210EP mist lubricant in MQL in contrast to dry machining conditions (Kajaria et al., 2012). Vazquez et al. conducted micro-milling experimental investigations on Ti-6Al-4V to analyze the effect of cutting fluid through computational fluid dynamics analysis. They observed that the best results on the basis of superior surface finish, tool life, accuracy, and geometrical shapes were observed when micro-milling under MQL conditions (Vazquez et al., 2014). Li et al. conducted micro-milling studies while machining on SKD61 steels having a hardness of 38 HRC under MQL conditions. A vegetable-based oil was used as the base cutting oil, and the coolant was supplied at three levels of flow rates. The authors observed an overall reduction in tool flank wear lengths by about 60% under all levels of cutting conditions in the case of MQL in contrast to dry machining conditions (Li & Chou, 2010). Figure 7.4 illustrates a graph representing tool flank wear and variation in surface roughness characteristics at various machining speeds. Moreover, it was observed that the surface roughness values were always less than 0.2 μm for all levels in the MQL condition, while it

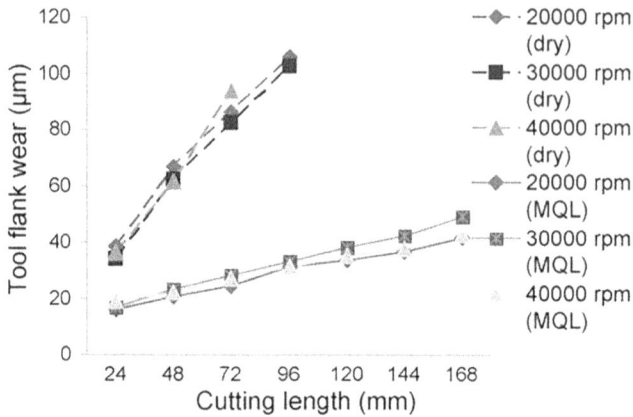

(a)

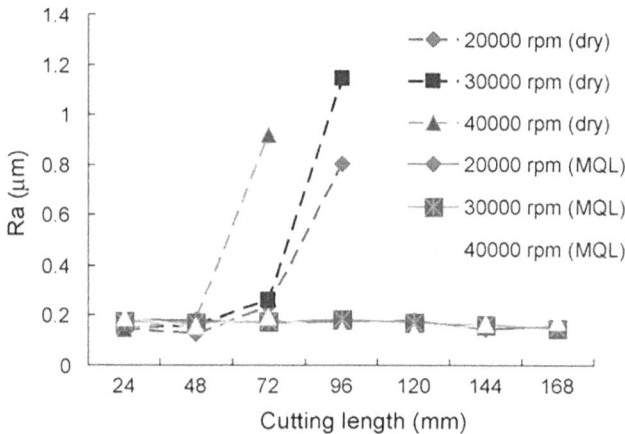

(b)

FIGURE 7.4 Graph representing (a) tool wear and (b) surface roughness at various machining speeds and lubrication conditions (Li & Chou, 2010).

reached as high as 1.1 μm in some cases of dry machining conditions. It was also inferred from the studies that a small amount of cutting oil at 1.88 mL/h was essential in the case of lubrication at near-micro-milling operations, while the pure air alone may not enhance the tool life significantly.

Vazquez et al. analyzed the influence of MQL, which used vegetable oil with viscosity 34cSt as the cutting fluid on tool life, quality of surface and surface burr generation during the machining of Ti-6Al-4V alloy. The authors observed a significant reduction in surface roughness and machining temperatures by 60% and 50% as compared to dry machining conditions (Vazquez et al., 2015). Figure 7.5 displays pictures of burr formation during various lubrication/cooling methods. Moreover, the best results in terms of improved tool life and reduced

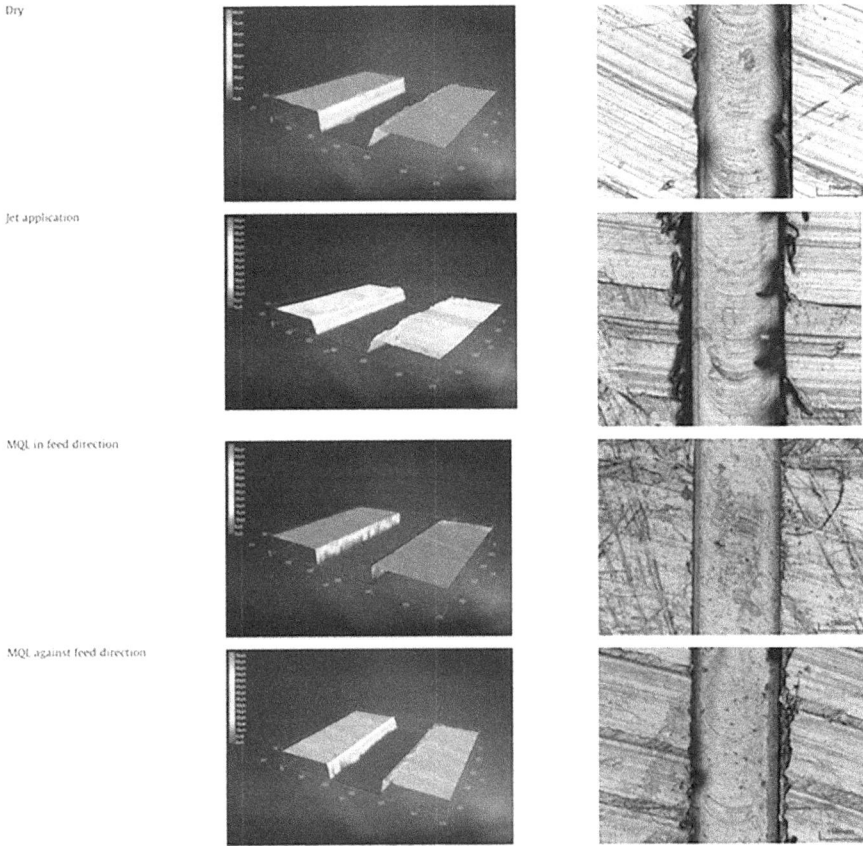

FIGURE 7.5 Pictures of burr formation according to different lubrication/cooling methods (Vazquez et al., 2015).

burr formation were observed while machining the alloy along the feed direction. Roushan et al. attempted to evaluate the impact of tool coatings as well as nanoparticle-incorporated MQL during micro-machining of Ti-6Al-4V. The authors prepared water-based CuO nanofluids as the minimum quantity lubricant. The nanofluid was prepared by adding CuO nanoparticle concentrations in deionized water and then by adding sodium dodecyl sulfate surfactant to it to the required proportions for uniform dispersion of the fluid. It was observed that a significantly lower percentage of reduction in tool edge diameter and BUE formation was noted in the case of uncoated WC endmill cutter under 1 vol% CuO nanofluid-MQL as a result of its superior heat dissipation features from the machining zone at higher CuO concentration nanofluids (Roushan et al., 2021). Figure 7.6 represents SEM images of various tools in MQL condition after 450 m of cutting length. Moreover, a higher surface finish was observed in the case of AlTiN-coated WC micro-end mill tools under 0.25 vol% CuO nanofluid MQL cutting environments.

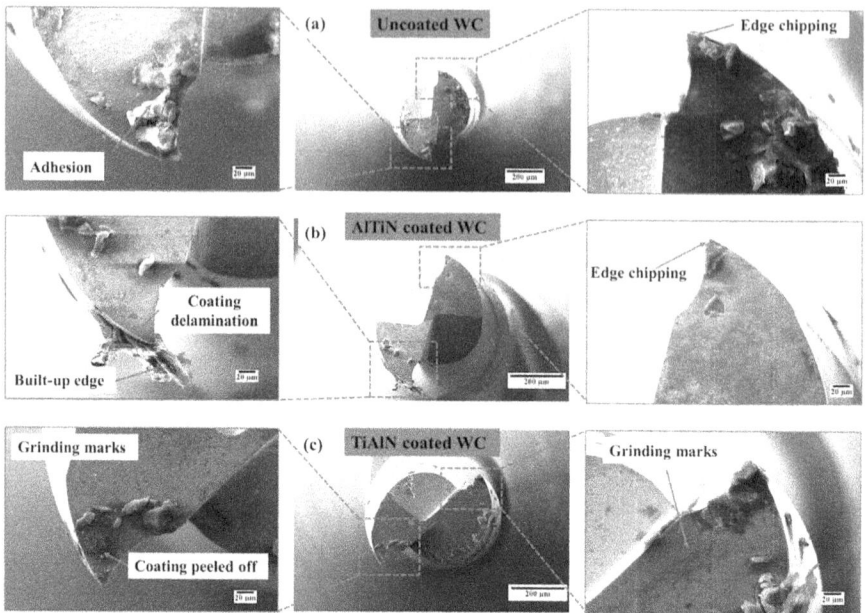

FIGURE 7.6 SEM images: (a) uncoated, (b) AlTiN-coated, and (c) TiAlN-coated WC micro end-mills in pure MQL condition after 450 mm cutting length. (Roushan et al., 2021).

Kim et al. conducted experimental investigations to analyze the effect of nanofluid-assisted MQL conditions while machining Ti-6Al-4V alloy. It was concluded from the experimental results that nanofluid-assisted MQL that uses nanodiamond particles as the nano-additives were significantly more effective in reducing the burr formation and surface roughness while performing micro-end milling operation on the alloy (Kim et al., 2014). The application of a graphene-based nanofluid during the micro-milling of SKH-9 steel was performed by Huang et al. Graphene, due to its higher strength and thermal conductivity combined with its superior lubricating properties, was selected as the nanoparticle to be supplied with compressed air during the micro-milling of the alloy. The results showed that the tool cutter wear and burr formation was minimal in the case of nanofluid-assisted MQL condition as the nanographene particles, due to their superior thermal conductivity characteristics, would carry away the heat experienced along the machining interface (Huang et al., 2020).

7.2.2 MICRO-DRILLING

Azim et al. explored the application of PVD-based TiAlN coating while micro-drilling of Incoloy 825 under MQL-based conditions. They observed that although the coating does not have a significant influence on machinability characteristics at low feed conditions, it was found to be effective in the presence of MQL-assisted conditions at higher feeds and spindle speed conditions. Based on the results, they concluded that a relatively poor achievement of the coated drill bits in terms of superior oversize error and deformed layer thickness at a feed rate of 1 μm/sec and a spindle speed of 20,000 rpm was attributed to the size effect phenomenon (Azim et al., 2020). Percin et al. conducted experimental studies to analyze the impacts of different lubrication conditions like dry cooling, flood cooling, MQL cooling, and cryogenic cooling in micro-drilling of Ti-6Al-4V alloy. The experiments were conducted using uncoated WC drill bits at various conditions of speed and feed rates. The authors noted that while cryogenic cooling resulted in a significant increase in the level of thrust forces, the MQL technique provided superior engagement torque amplitude (Perçin et al., 2016). Khanafer et al. studied the influence of various lubrication conditions like flood cooling, pure-MQL cooling, and nanofluid-assisted MQL cooling during micro-drilling of Inconel 718. Aluminium oxide nanoparticles were used as the nano-additives for preparing the nanofluid due to their excellent tribological properties and antitoxic characteristics. The nano-additives were then supplied with the vegetable oil–based MQL mist to the machining interface. From the results, it was inferred that when Inconel 718 was subjected to micro-drilling under nanofluid-assisted MQL conditions, there occurred a significant decrease in tool flank wear and deterioration at outer corners as compared to flood and MQL cooling conditions. This may be attributed to the fact that the nano-additives during MQL cooling can easily infiltrate the machining zone more significantly, whereby the nanosized particles can facilitate a ball-bearing mechanism between the tool-work interface, which significantly minimizes the frictional forces (Khanafer et al., 2020). Figure 7.7 shows the SEM images of tools under various lubrication conditions. In addition to that, the utilization of nanofluid-MQL enhanced the drilling quality and significantly reduced the surface burr generation along the drilled holes.

Nam et al. analyzed the micro-machinability of the Ti-6Al-4V alloy under nanofluid-assisted MQL conditions. For the preparation of nanofluid, nano-diamond particles with sizes 35 nm and 80 nm were selected with vegetable oil as the cutting fluid. It was noted from the experimental results that nano-sized diamond particles with a size of 35 μm and a higher weight concentration of 0.4% wt, and a lower feed rate of 10 mm/min were beneficial for enhancing the micro-machinability of Ti-6Al-4V in terms of the thrust forces and drilling torques combined with improved tool life and hole quality while machining under nanoparticle-assisted MQL conditions (Nam & Lee, 2018). Figure 7.8 shows the edge radius effect and circularity error measured while micro-drilling at various lubrication conditions, respectively. These significant improvements in results may be attributed to the improved ball-bearing effect, which could result in enhanced lubrication at the drilling interface that could effectively reduce the chip adhering to the drill chisel area. Nam et al. conducted an experimental investigation to perform optimization of nanofluid-assisted MQL micro-drilling process for grade 5 titanium alloy using response surface methodology (RSM). While examining the ANOVA results, they concluded that the diameter of the drill and spindle speed were the most significant factors that affected the edge radius and drill torque as well as thrust force (Nam et al., 2018). On the other hand, self-interactions of feed rate and weight concentrations were found to be statistically significant. Similarly, the optimization of environmentally friendly nanofluid-assisted MQL in micro-drilling operation with nanodiamond particles incorporated in base fluids of paraffin and vegetable oil was performed by Nam et al. based on RSM and GA techniques (Nam et al., 2015).

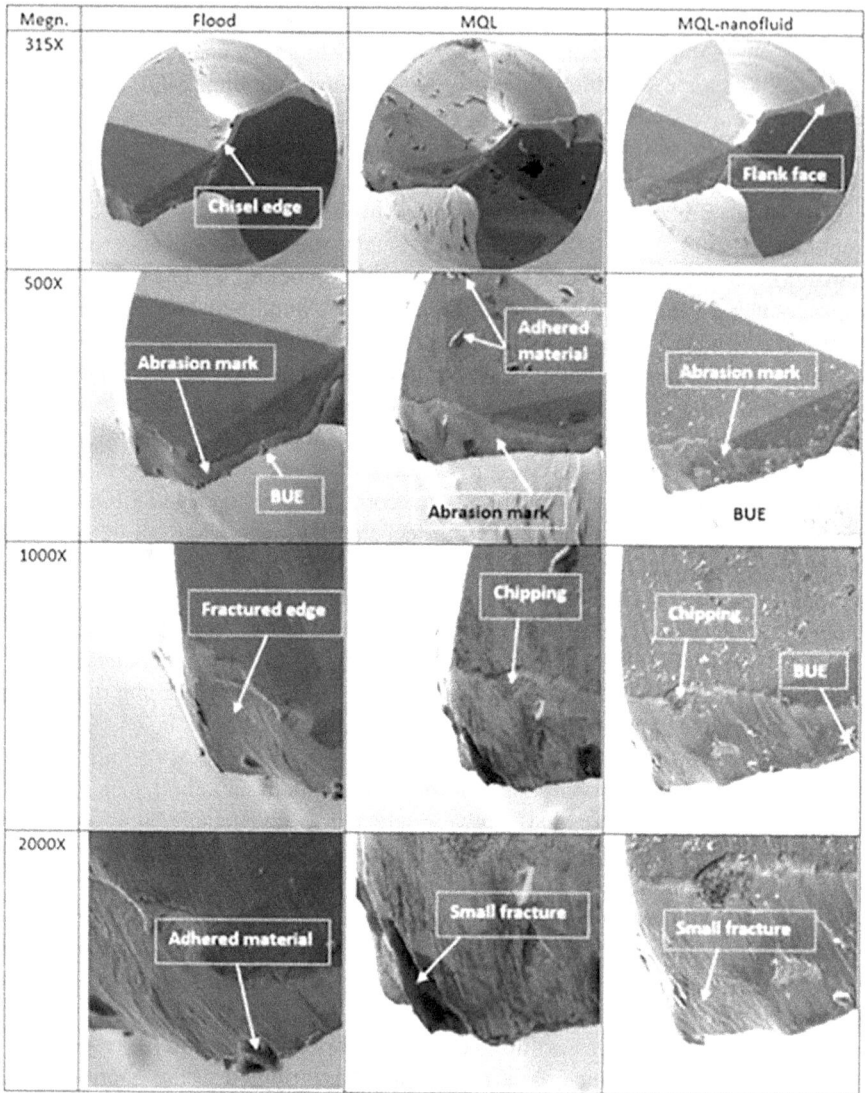

FIGURE 7.7 SEM images of tool flank face after drilling 30 holes under various conditions of flood cooling, MQL, and nanofluid-MQL cooling (Khanafer et al., 2020).

As a part of multiple response optimization, while average torque, as well as thrust forces, were curtailed, the material removal rate was maximized. Coefficients of determination (R^2) were observed to be greater than 93% when regression equations were validated by ANOVA. It was also observed that diameter had a greater significance in contrast to feed rates and depth of cut values. The optimal combination was found to have a drill diameter of 0.5 mm, a spindle speed of about 59,000 rpm, and a volumetric concentration of nanofluid by about 2%.

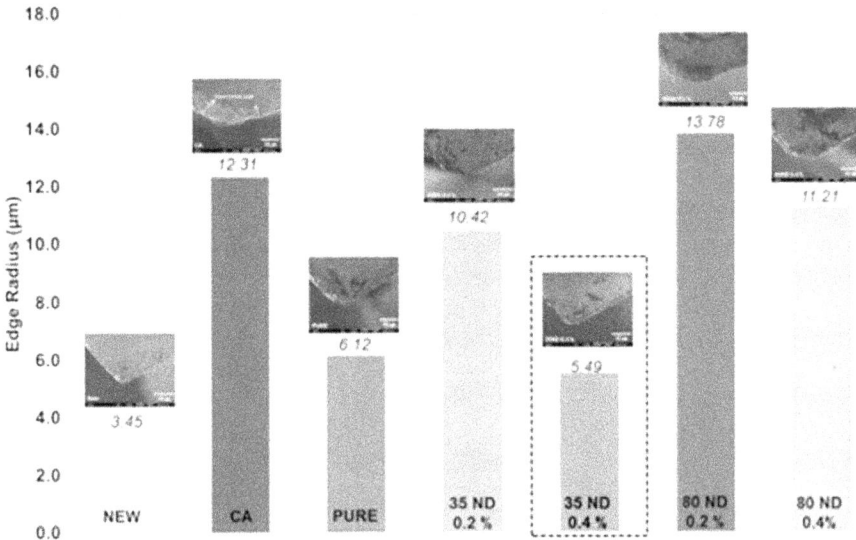

(a)

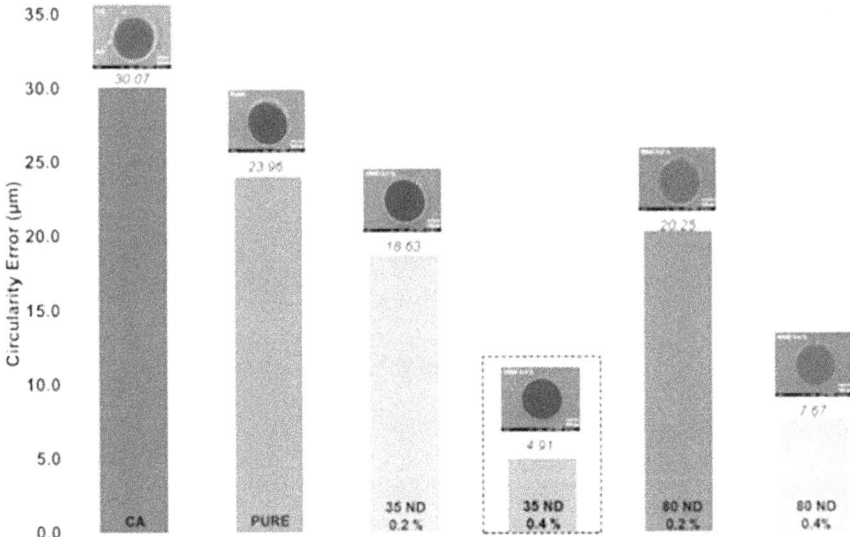

(b)

FIGURE 7.8 Graphs showing (a) drill bit edge radius and (b) circularity error measured during various lubrication conditions while micro-drilling Ti-6Al-4V (Nam & Lee, 2018).

In addition to that, a feed rate of 5 mm/min was noted to be the optimal value when considering vegetable oil, while a feed rate of 10.6 mm/min was found to be the optimal value when considering paraffin oil as the base cutting fluid. The influence of nanofluid-assisted MQL in micro-drilling with the incorporation of nanodiamond components was analyzed by Nam et al. Vegetable oil and paraffin oil were used as the base fluid in MQL and the experiments were performed at various concentrations of nanoparticles. The authors were of the view that the micro-drills seldom experienced failure after performing 150 hole indentations, which may be ascribed to their improved cooling and lubrication phenomenon that minimized the chip adhering tendency to the drill bits (Nam et al., 2011). Moreover, the presence of nanodiamond components in MQL reduced the magnitude of the thrust forces and torques. Figure 7.9 shows the average torque and average thrust force experienced during

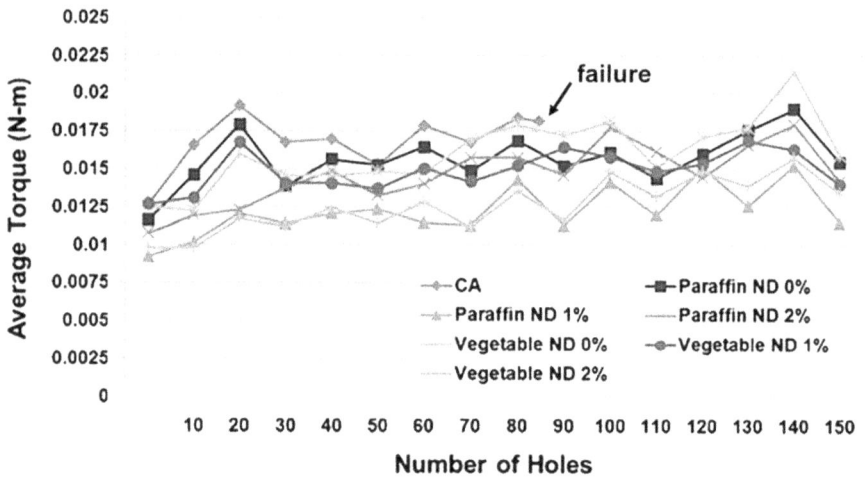

(a)

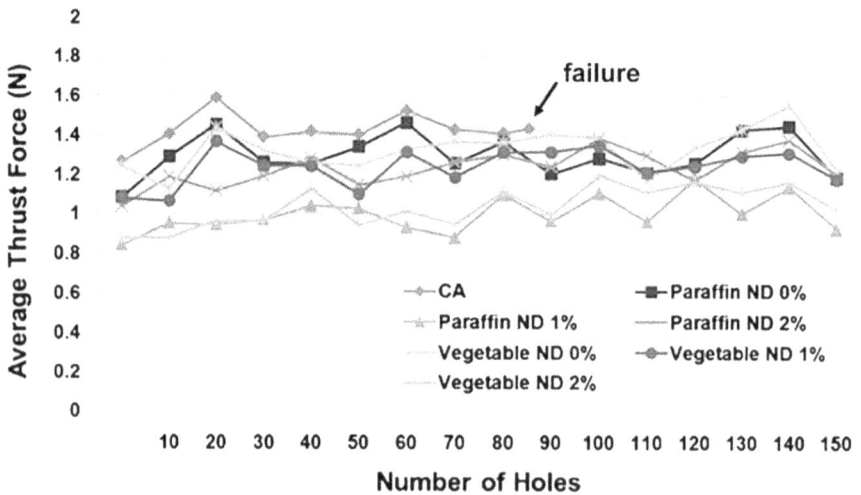

(b)

FIGURE 7.9   (a) Average drilling torques, (b) average thrust force vs. the holes measured at various lubrication conditions (Nam et al., 2011).

micro-drilling at various lubrication conditions. From an overall perspective, nano-fluid prepared by using paraffin oil as the base cutting fluid in MQL performed more significantly than vegetable-based oil. When considering paraffin-based oil, the addition of 1 vol% of nano-diamond components were found to give superior results in terms of reduced torques and forces in contrast to 2 vol% concentration. However, while considering vegetable-based oil, results were vice versa as the higher concentration of nanodiamond components yielded better significant results. The quality of the micro-holes also improved considerably while machining with nanofluid-assisted MQL condition due to its enhanced lubrication properties. Huang et al. presented a novel nanofluid-assisted MQL technique for analyzing the machinability characteristics of 7075-T6 aluminium alloy during its micro-drilling operation. Various concentrations of nanodiamond were added into a vegetable-based cutting oil as part of the MQL mixture. Taguchi design approach was used to analyze the optimal combinations of nanofluid-assisted MQL parameters. Reduced occurrences of rapid failure of the drill bits were observed under nanofluid-assisted MQL conditions. The optimal parameter combination was observed to be at air compressor pressure (2 bar), a distance of nozzle from machining zone (15 mm), and concentration of nanodiamond (2% in wt) (Huang et al., 2016).

A significant reduction across the machining temperature noted in nanofluid-assisted MQL conditions may be ascribed to the nanodiamond particle's improved thermal conductivity, which in turn significantly improved the overall heat dissipation characteristics. In addition to that, an improvement in overall hole quality and considerably reduced burr formation across the holes were observed during micro-drilling of the aluminium alloy under nanofluid-assisted MQL conditions. Figure 7.10 displays the average drilling temperature experiences across the tool faces at various lubrication conditions. Figure 7.11 and Figure 7.12 represent the SEM images of the tool rake face and the machined hole at various lubrication conditions, respectively.

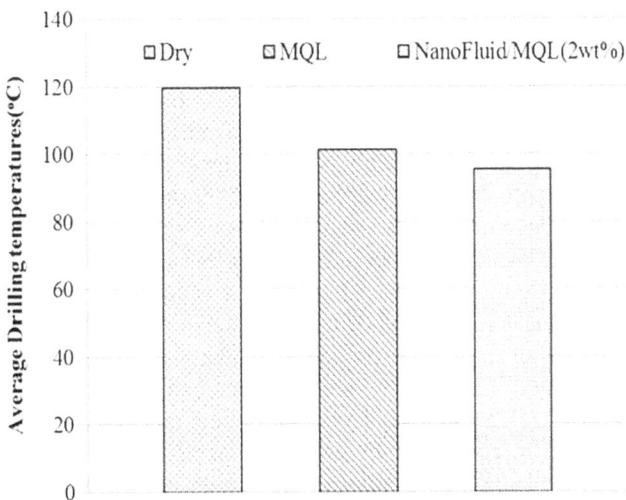

FIGURE 7.10 Average drilling temperatures experienced across the drill rake face at various lubrication conditions (Huang et al., 2016).

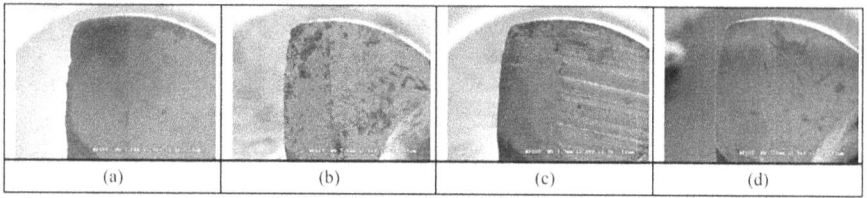

FIGURE 7.11 SEM images of rake faces of micro-drills at various lubrication conditions (Huang et al., 2016).

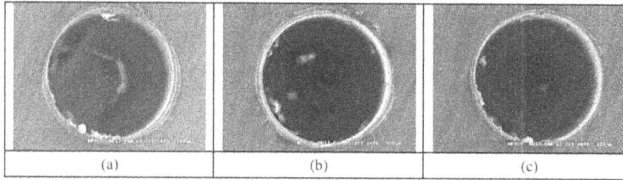

FIGURE 7.12 SEM images of holes machined using tools at various lubrication conditions (Huang et al., 2016).

7.2.3 Micro-turning

Oliveira et al. performed micro-machining experiments on Ti-6Al-4V to analyze the impact of various lubrication conditions like dry cooling, cold air, and MQL cooling on its machinability characteristics. The experiments were conducted using a ceramic tool and a TiN-coated carbide micro-turning insert. The authors concluded that the highest surface finish was achieved with a carbide tool having a chip breaker at lower feed rates and depth of cut at dry cutting. However, the tool geometry and the use of MQL conditions were more significant in causing hardening in the micro-scale at the material surface (de Oliveira et al., 2017).

Elias et al. developed micro-textures on the flank face of a micro-turning insert using the micro-indentation technique and analyzed its impact on tool life characteristics, machining forces, machining temperatures, and surface roughness of the specimen at various lubrications of dry and MQL during micro-turning of Ti-6Al-4V alloy. Micro-indentations with a square pyramid shape were developed on the micro-insert using a Vickers microhardness tester by supplying load in the range of 50 gf to 2,000 gf. Compressed air mixed with a vegetable-based cutting oil was sprayed across the machining interface with an MQL setup. The authors reported an overall drop in cutting forces, the roughness of the specimen, machining temperature and tool wear by 21%, 6%, 7%, and 19%, respectively, as compared to that of the plain tool inserts. The authors were of the view that the presence of textures along the tool face in combination with MQL cooling, helped in curtailing the frictional forces and cutting temperature significantly in contrast to normal turning conditions (Elias et al., 2021).

Filho et al. attempted a study on online monitoring in the micro-turning process to analyze the impacts of feed rates, cooling type, depth of cuts and type of tools on

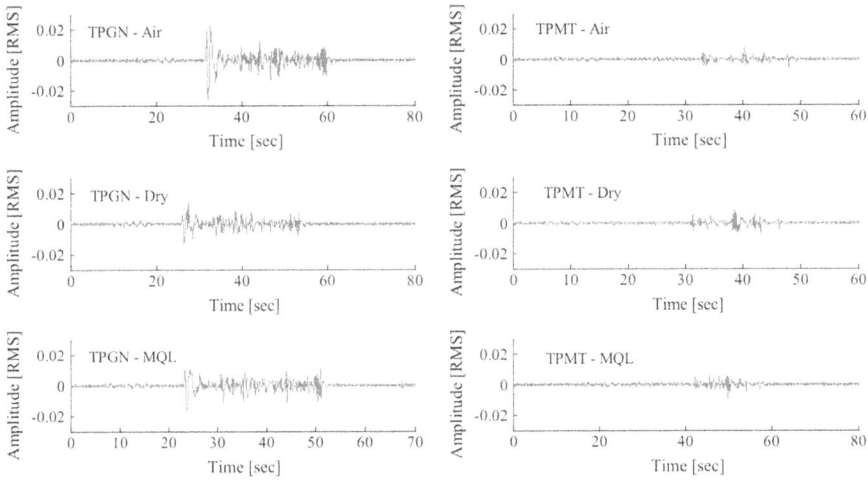

FIGURE 7.13 AE$_{RMS}$ signal; TPGN tool; (a) feed rate of 50 µm/rev, depth of cut of 100 µm, and TPMT tool, (b) feed rate of 120 µm/rev, depth of cut 500 µm for various lubrication conditions (Filho et al., 2019).

response acoustic emission signals. The experiments were conducted under dry cutting, air cooling, and MQL conditions using the TPGN tool and TPMT tool. The authors observed that the feed rate was the most significant factor that influenced the surface roughness values (Filho et al., 2019). Figure 7.13 shows the acoustic signals generated for different types of tools at various machining conditions. Moreover, micro-scale hardening at the material surface was influenced based on tool geometry and MQL condition.

7.2.4 MICRO-GRINDING

Ming-Li et al. performed micro-grinding experiments to examine the influence of MQL on tool life characteristics and ground surface roughness. A micro-grinding tool having 600 µm diameter and grain size 200 with a diamond abrasive type was utilized. SK3 was used as the workpiece specimen and a vegetable-based fluid was used as the cutting fluid. From the experiments conducted, it was observed that minimal surface roughness with reduced burn marks was noted in the case of MQL assisted condition due to the limited accumulation of chip on the grinding tool in comparison to micro-grinding in dry machining conditions (Li & Lin, 2012). Figure 7.14 shows the material removal rates and surface roughness values for different cooling conditions. Moreover, the tool life was observed to be seven times higher than at dry micro-grinding and about five times higher than that of micro-grinding at air cooling conditions, respectively. In addition to that, an oil flow rate of 1.88 mL/h and airflow of 25 L/min were found to be the recommended combination of parameters in MQL conditions.

Liu et al. investigated the influence of a combined effect of atmospheric pressure plasma jet cooling (APPJ) and MQL during micro-grinding of a quenched GCr15

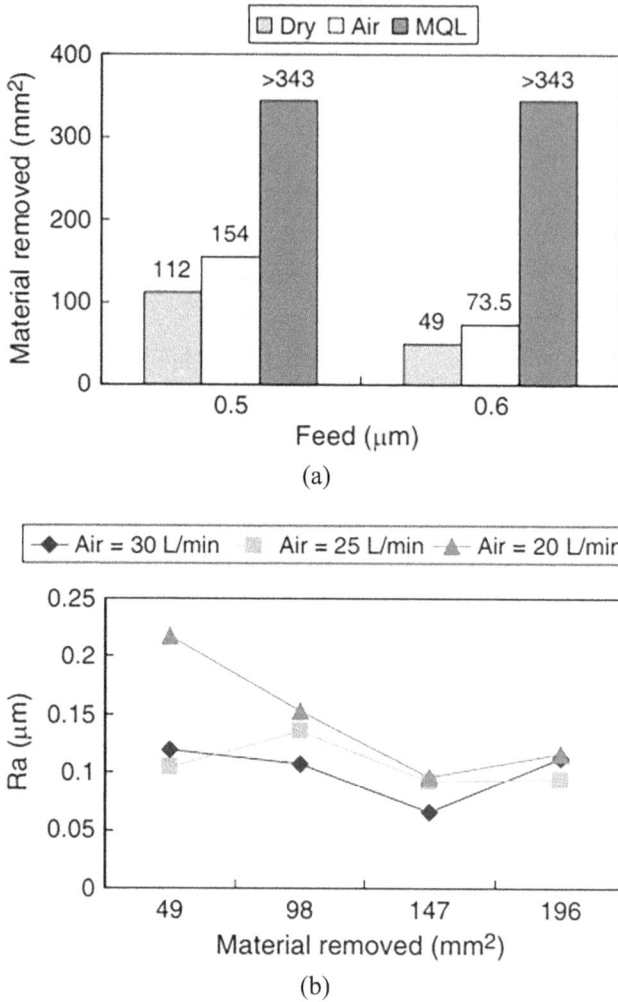

FIGURE 7.14   Graphs showing (a) tool life and (b) surface roughness for different air flow rates (Li & Lin, 2012).

specimen. Micro-grinding investigations were conducted under different lubrication conditions like dry grinding, nitrogen-assisted grinding, MQL-assisted and APPJ-assisted grinding, as well as a combination of APPJ and MQL-assisted grinding conditions. It was inferred from the experimental investigations that the combination of APPJ and MQL helped in minimizing the machining temperature by around 47% as compared to dry micro-grinding conditions, which may be attributed to the hydrophilization phenomenon of APPJ and phase transition during the MQL cooling period (Liu et al., 2020).

Moreover, the improved surface quality of the specimen and reduction in overall cutting forces by 82% and 48% were observed in contrast to dry and MQL

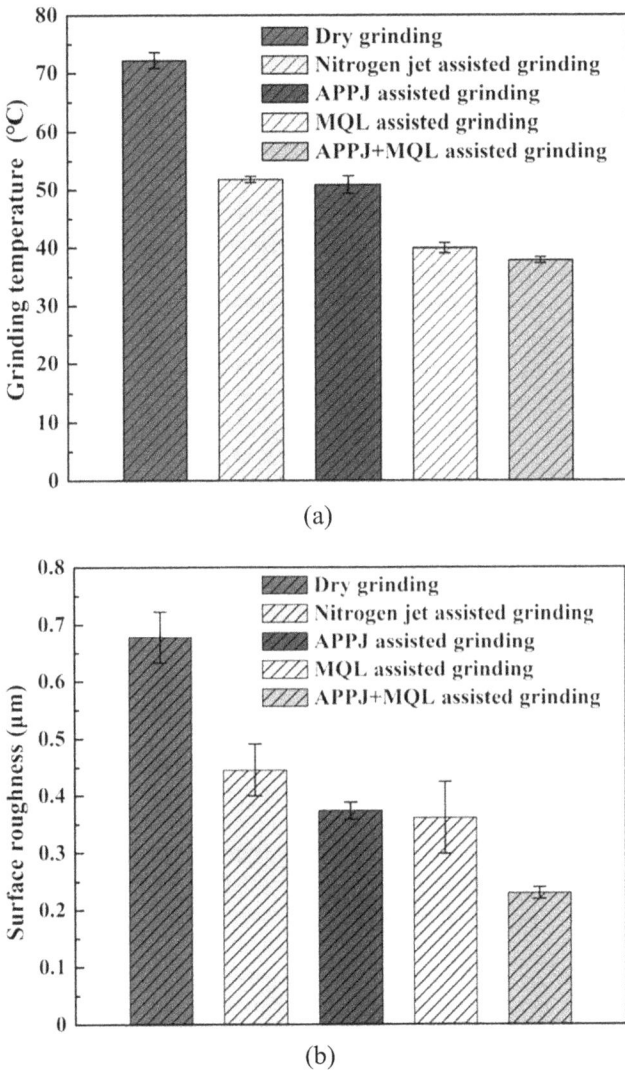

FIGURE 7.15 Graphs shows (a) grinding temperature (°C) and (b) surface roughness values at various lubrication conditions (Liu et al., 2020).

micro-grinding conditions, respectively. Figure 7.15 indicates graphs showing grinding temperature and surface roughness values generated during various lubrication conditions. In addition to that, while micro-grinding in dry conditions yielded a specific grinding energy of 738 J/mm^3, the combination of APPJ and MQL yielded a specific cutting energy of only 129 J/mm^3, which shows the overall improvement in the micro-grinding process during APPJ and MQL-assisted conditions.

Lee et al. analyzed the significance of nanofluid-assisted MQL in the mesoscale grinding process. The nanofluids were prepared by using nanodiamond components in paraffin oil. During experimental investigations, it was observed that tangential as well as normal forces decreased by 30.3% and 33.2%, respectively, as compared to dry air grinding conditions. Moreover, it was also inferred from the experimental results that nanofluid prepared by the incorporation of nano-sized diamond particles with a 30 nm size reduced the ground surface roughness by about 64% compared to dry air grinding conditions (Lee et al., 2010).

Lee et al. examined the influence of nanoparticle-assisted MQL during micro-grinding operations performed in a miniaturized grinding tool system. Al_2O_3 and nanodiamond components were selected based on their thermal conductivities and improved hardness characteristics, while paraffin oil was selected as the base cutting fluid. It was observed from the experimental results that while the incorporation of nanoparticles in MQL helped in minimizing the overall cutting forces significantly, nanodiamond components were deemed to be more effective in reducing the grinding forces more effectively as compared to Al_2O_3 particles as a part of their enhanced lubrication effects. Figure 7.16 displays graphs showing normal and

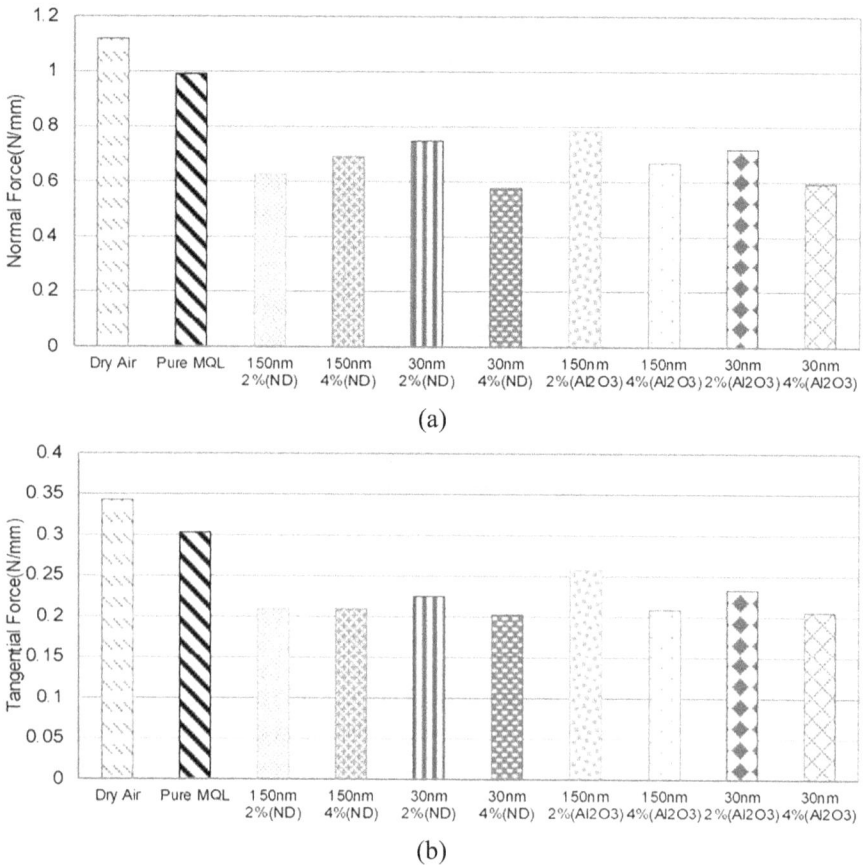

(a)

(b)

FIGURE 7.16 Graphs showing (a) normal Force and (b) tangential forces developed in various lubrication-assisted micro-grinding conditions (Lee et al., 2012).

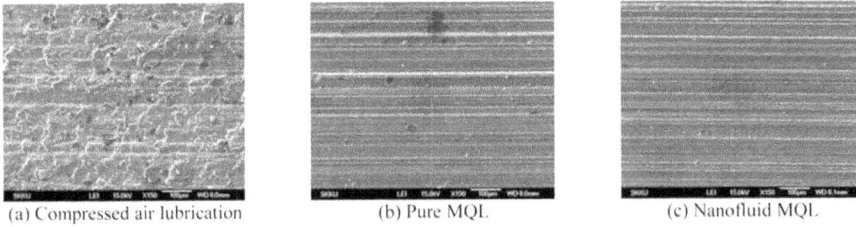

(a) Compressed air lubrication   (b) Pure MQL   (c) Nanofluid MQL

FIGURE 7.17 SEM images of the machined ground surfaces at various lubrication conditions (Lee et al., 2015).

tangential forces developed during micro-grinding. Meanwhile, nanosized Al_2O_3 particles were observed to be more effective in minimizing the overall surface roughness compared to nanodiamond particles due to the reduced hardness levels of the Al_2O_3 particles (Lee et al., 2012).

Lee et al. examined the thermal characteristics of the micro-grinding process with a miniaturized micro-grinding system with the assistance of the nanofluid MQL technique with analytical as well as experimental analysis. Paraffin oil was used as the base cutting oil, while nanodiamond particles were incorporated into the cutting fluid. It was noted from experimental results that the reduction in tangential forces was observed to be 16.7% in nanofluid-MQL than that of the pure MQL condition. Figure 7.17 shows SEM images of ground surfaces after micro-grinding at various lubrication conditions. Moreover, minimized grinding temperatures were also observed in the case of nanofluid-assisted MQL condition, which may be attributed to the ball-bearing phenomenon of the nanodiamond particles at the machining interface (Lee et al., 2015).

While inspecting the surface quality of the ground specimen, a considerable amount of plough marks and recast chips were observed along the ground surface in the case of dry and pure-MQL grinding conditions due to its insignificant cooling and lubrication conditions. In addition to that, temperatures estimated from the numerical model yielded good agreement on experimental validation of the results.

7.3 CONCLUSION

This chapter discusses some of the various trends carried out by researchers in the field of micro-machining with the application of minimum quantity lubrication (MQL) techniques. The works pertained to the usage of MQL and its recent advancements in various micro-machining operations like micro-milling, micro-drilling, micro-turning, and micro-grinding with an eye on overall improvement in machinability characteristics were discussed in detail. Some of the major conclusions inferred from the study are as listed below:

- MQL had a favorable effect on the machinability characteristics of the alloy due to its ability to infiltrate across the machining interface and provide significant cooling and lubrication effects, unlike conventional cooling techniques.

- The use of biodegradable cutting oils like vegetable-based cutting fluids in MQL, implemented as a part of sustainable and environment-friendly manufacturing techniques, improved the overall performance of the cutting tools in ensuring minimal tool failure rates while also easing concerns over the disposal of the cutting fluids.
- Incorporation of nanosized particles with a higher thermal conductivity and lower coefficient of friction into the MQL mixture helped in improving the overall machinability by enhancing its cooling rates while also ensuring a reduced contact pressure at the machining interface due to its improved load bearing tendency.
- As a part of future scope, the use of MQL in combination with tools subjected to cryogenic treatment techniques prior to actual micro-machining operations can be carried out. While the deep cryogenic tool treatment technique enhances the overall wear resistance of the tool due to its improved overall electrical and thermal conductivity, the application of MQL with the incorporation of nanoparticles may help in relieving the contact pressure experienced at the machining interface.

REFERENCES

Anand, K. N., & Mathew, J. (2020). Evaluation of size effect and improvement in surface characteristics using sunflower oil-based MQL for sustainable micro-endmilling of Inconel 718. *Journal of Brazilian Society of Mechanical Sciences and Engineering*, 42, 156. 10.1007/s40430-020-2239-0

Anand, K. N., & Mathew, J. (2021). Size effect and micro endmilling performance while sustainable machining on Inconel 718. *Materials and Manufacturing Processes*, 36 (6), 668–676. 10.1080/10426914.2020.1854465

Azim, S., Gangopadhyay, S., Mahapatra, S. S., Mittal, R. K., & Singh, R. K. (2020). Role of PVD coating on wear and surface integrity during environment-friendly micro-drilling of Ni-based superalloy. *Journal of Cleaner Production*, 272, 122741.

de Oliveira, J. A., Ribeiro Filho, S. L. M., Lauro, C. H., & Brandão, L. C. (2017). Analysis of the micro turning process in the Ti-6Al-4V titanium alloy. *International Journal of Advanced Manufacturing Technology*, 92 (9–12), 4009–4016.

Dhar, N. R., Islam, M. W., Islam, S., & Mithu, M. A. H. (2006). The influence of minimum quantity of lubrication (MQL) on cutting temperature, chip and dimensional accuracy in turning AISI-1040 steel. *Journal of Materials Processing Technology*, 171(1), 93–99.

Elias, J. V., Venkatesh N, P., Lawrence K, D., & Mathew, J. (2021). Tool texturing for micro-turning applications–an approach using mechanical micro indentation. *Materials and Manufacturing Processes*, 36 (1), 84–93.

Hassanpour, H., Sadeghi, M. H., Rezaei, H., & Rasti, A. (2016). Experimental Study of Cutting Force, Microhardness, Surface Roughness, and Burr Size on Micromilling of Ti6Al4V in Minimum Quantity Lubrication. *Materials and Manufacturing Processes*, 31 (13), 1654–1662.

Huang, W. T., Wu, D. H., & Chen, J. T. (2016). Robust design of using nanofluid/MQL in micro-drilling. *International Journal of Advanced Manufacturing Technology*, 85 (9–12), 2155–2161.

Huang, W. T., Chou, F. I., Tsai, J. T., & Chou, J. H. (2020). Application of Graphene Nanofluid/Ultrasonic Atomization MQL System in Micromilling and Development of

Optimal Predictive Model for SKH-9 High-Speed Steel Using Fuzzy-Logic-Based Multi-objective Design. *International Journal of Fuzzy Systems*, 22 (7), 2101–2118.

Joshy, J., & Kuriachen, B. (2023). Influence of post-processing and build orientation on the micro-machinability and chip formation during micro-drilling of L-PBF AlSi10Mg. *CIRP Journal of Manufacturing Science and Technology*, 45, 35–48.

Kajaria, S., Chittipolu, S., Adera, S., & Hung, W. N. (2012). Micromilling in minimum quantity lubrication. *Machining Science and Technology*, 16 (4), 524–546.

Khanafer, K., Eltaggaz, A., Deiab, I., Agarwal, H., & Abdul-latif, A. (2020). Toward sustainable micro-drilling of Inconel 718 superalloy using MQL-Nanofluid. *International Journal of Advanced Manufacturing Technology*, 107 (7–8), 3459–3469.

Kim, D. H., Lee, P.-H., Kim, J. S., Moon, H., & Lee, S. W. (2014). Experimental Study on Micro End-Milling Process of Ti-6Al-4V Using Nanofluid Minimum Quantity Lubrication (MQL). Proc. of International Manufacturing Science and Engineering Conference on American Society of Mechanical Engineers.

Lee, P. H., Nam, T. S., Li, C., & Lee, S. W. (2010). Environmentally-friendly nano-fluid minimum quantity lubrication (MQL) meso-scale grinding process using nano-diamond particles. *Proc. - 2010 Int. Conf. Manuf. Autom. ICMA 2010*, 44–49.

Lee, P. H., Nam, J. S., Li, C., & Lee, S. W. (2012). An experimental study on micro-grinding process with nanofluid minimum quantity lubrication (MQL). *International Journal of Precision Engineering and Manufacturing*, 13 (3), 331–338.

Lee, P. H., Lee, S. W., Lim, S. H., Lee, S. H., Ko, H. S., & Shin, S. W. (2015). A study on thermal characteristics of micro-scale grinding process using nanofluid minimum quantity lubrication (MQL). *International Journal of Precision Engineering and Manufacturing*, 16 (9), 1899–1909.

Li, K. M., & Chou, S. Y. (2010). Experimental evaluation of minimum quantity lubrication in near micro-milling. *Journal of Materials Processing Technology*, 210(15), 2163–2170.

Li, K. M., & Lin, C. P. (2012). Study on minimum quantity lubrication in micro-grinding. *International Journal of Advanced Manufacturing Technology*, 62 (1–4), 99–105.

Liu, J. *et al.* (2020). Atmospheric pressure plasma jet and minimum quantity lubrication assisted micro-grinding of quenched GCr15. *International Journal of Advanced Manufacturing Technology*, 106 (1–2), 191–199.

Mahesh, K., Philip, J. T., Joshi, S. N., & Kuriachen, B. (2021). Machinability of Inconel 718: A critical review on the impact of cutting temperatures. *Materials and Manufacturing Processes*, 36(7), 753–791.

Manikandan, N., Arulkirubakaran, D., Palanisamy, D., & Raju, R. (2019). Influence of wire-EDM textured conventional tungsten carbide inserts in machining of aerospace materials (Ti–6Al–4V alloy). *Materials and Manufacturing Processes*, 34 (1), 103–111.

Masuzawa, T. (2000). State of the art of micromachining. *CIRP Ann. - Manuf. Technol.*, 49(2), 473–488.

Nam, J., & Lee, S. W. (2018). Machinability of titanium alloy (Ti-6Al-4V) in environmentally-friendly micro-drilling process with nanofluid minimum quantity lubrication using nanodiamond particles. *International Journal of Precision Engineering and Manufacturing - Green Technology*, 5 (1), 29–35.

Nam, J., Kim, J. W., Kim, J. S., Lee, J., & Lee, S. W. (2018). Parametric analysis and optimization of nanofluid minimum quantity lubrication micro-drilling process for titanium alloy (Ti-6Al-4V) using response surface methodology and desirability function. *Procedia Manuf.*, 26, 403–414.

Nam, J. S., Lee, P. H., & Lee, S. W. (2011). Experimental characterization of micro-drilling process using nanofluid minimum quantity lubrication. *International Journal of Machine Tools and Manufacture*, 51 (7–8), 649–652.

Nam, J. S., Kim, D. H., Chung, H., & Lee, S. W. (2015). Optimization of environmentally benign micro-drilling process with nanofluid minimum quantity lubrication using response surface methodology and genetic algorithm. *Journal of Cleaner Production*, 102, 428–436.

Öndin, O., Kıvak, T., Sarıkaya, M., & Yıldırım, Ç. V. (2020). Investigation of the influence of MWCNTs mixed nanofluid on the machinability characteristics of PH 13-8 Mo stainless steel. *Tribology International*, 148, 106323.

Patole, P. B., & Kulkarni, V. V. (2018). Prediction of surface roughness and cutting force under MQL turning of AISI 4340 with nano fluid by using response surface methodology. *Manufacturing Review*, 5, 1–12. 10.1051/mfreview/2018002

Perçin, M., Aslantas, K., Ucun, I., Kaynak, Y., & Çicek, A. (2016). Micro-drilling of Ti-6Al-4V alloy: The effects of cooling/lubricating. *Precision Engineering*, 45, 450–462.

Ribeiro Filho, S. L. M., de Oliveira, J. A., Lauro, C. H., & Brandão, L. C. (2019). Monitoring of microturning process using acoustic emission signals. *Journal of Brazilian Society of Mechanical Sciences and Engineering*, 41 (6), 1–11.

Roushan, A., Rao, U. S., Patra, K., & Sahoo, P. (2021). Performance evaluation of tool coatings and nanofluid MQL on the micro-machinability of Ti-6Al-4V. *Journal of Manufacturing Processes*, 73, 595–610.

Ucun, I., Aslantasx, K., & Bedir, F. (2015). The effect of minimum quantity lubrication and cryogenic pre-cooling on cutting performance in the micro milling of Inconel 718. *Proc. Inst. Mech. Eng. Part B J. Eng. Manuf.*, 229 (12), 2134–2143.

Vazquez, E., Kemmoku, D. T., Noritomi, P. Y., Da Silva, J. V. L., & Ciurana, J. (2014). Computer fluid dynamics analysis for efficient cooling and lubrication conditions in micromilling of Ti6Al4V alloy. *Materials and Manufacturing Processes*, 29, (11–12), 1494–1501.

Vazquez, E., Gomar, J., Ciurana, J., & Rodríguez, C. A. (2015). Analyzing effects of cooling and lubrication conditions in micromilling of Ti6Al4V. *Journal of Cleaner Production*, 87 (1), 906–913.

8 Minimum quantity lubrication (MQL) for different materials

8.1 ALUMINUM

Aluminum (Al) and its alloys find great applications in today's manufacturing sector due to their high strength-to-weight ratio, excellent corrosive resistance, and flexibility in fabrication. Machining of these alloys such as 6061 Al alloy, Al 7075, 6061-T6, 7076-T6, A356, AA7075-T6, 2024-T351, 6026-T9, and AA 2024 T3 through turning, milling, drilling, tapping, etc. leads to heat generation and thus high cutting temperatures are generated in the process. The MQL technique is one of the alternatives in machining to achieve lower cutting force and cutting temperature compared to conventional flood coolant machining. The current section focused on the effect of MQL on cutting force, cutting temperature, tool wear, tool life, surface roughness, and chip formation in various machining operations while machining Al alloys.

8.1.1 EFFECT OF MQL ON CUTTING FORCE

The resultant cutting force in a dry machining condition is very high due to adhesion of the workpiece material to the cutting tool as aluminum is a soft material. At the initial stages of machining when the cutting tool is sharp, the cutting force is low; however, the cutting force increases as the adhesion of the workpiece material progresses. Due to high pressure coolant/lubricant supply in MQL machining, the lubricant particles reach the contact area between the workpiece and the cutting tool, thereby reducing the heat generation due to friction. This reduction is heat generation helps to reduce temperature on the cutting tool, which reduces adhesion and in turn reduces frictional force.

Cutting force generation also depends on material composition as this may lead to different deformation mechanisms in machining. The MQL application in machining of Al6061 and Al7075 alloys resulted in no significant difference in cutting forces, but in machining of Al2024 alloy the cutting force was very high, especially at low and intermediate speeds due to different deformation manners. The type of liquid and flow rate of the liquid influences the cutting force in machining with the MQL application. The cutting force was higher at low flow rates and the cutting force was reduced as the flow rate of the MQL system increased. The increase in flow rate minimizes the coefficient of friction and increases chip flow velocity over the tool rake surface. The chip-tool contact length reduces with an increase in chip flow velocity and thus helps to reduce cutting forces. However, the flow rate has no

DOI: 10.1201/9781003328742-8

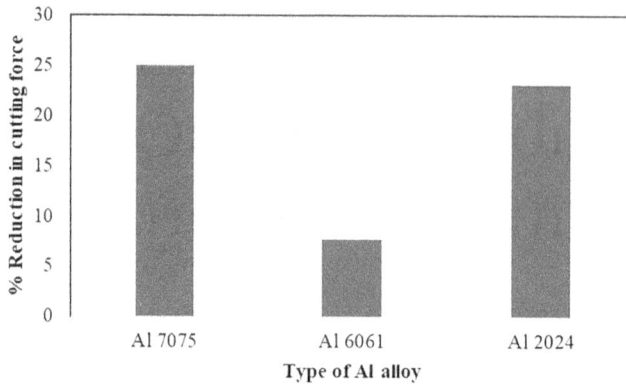

FIGURE 8.1 Percentage reduction in cutting force with MQL compared to dry cutting while machining different Al alloys.

influence on the cutting force at very high cutting speeds as the lubrication mode diminishes and most of the heat was evacuated by the chip (Khettabi et al., 2017) (Figure 8.1).

8.1.2 EFFECT OF MQL ON TEMPERATURE

The cutting temperature has a significant influence on the quality of the workpiece and tool life. The cutting temperature increases due to high shear strength of the chip in chip formation process (Junge et al., 2022). MQL fluids have less heat-removing abilities compared to water-based fluids; however, they have good wettability and lubricity. The penetrating ability of MQL oil mist helps in reducing the coefficient of friction between chip tool contact area and thus reduces the heat generation due to frictional forces. The reduction in heat generation helps to lower the cutting temperature in the machining process with MQL (Liu et al., 2010). Hence, the lubrication capacity and thermal conductivity of the lubricant used in MQL plays important role in temperature reduction in machining zone.

Some of the researchers developed a coolant supply system in which water droplets are covered with oil film (oil-on-water) and sprayed in the machining zone with a nozzle. Machining of aluminum alloy (JIS A6063) with the help of this system provided superior cooling performance compared to a dry cutting condition. This method resulted in a minimum cutting temperature due to sensible and latent heat properties of oil on water, and effectively decreased the percentage of the defective parts in machining. The use of oil on water in machining reduced almost 50% the cutting temperature (Yoshimura et al., 2005). In another case, drilling of Al alloy 7075 under the MQL environment using boron oil-water emulsion achieved a temperature reduction of 40% by a chilling effect of the water-spreading phenomena (Kilickap et al., 2011). The use of nanofluid as a coolant/lubricant in MQL while machining of Al alloys helped to reduce the cutting temperature. The nanofluid performance depends on the concentration and operating temperature in correlation with the lubrication and cooling capacity of the nanofluid (Figure 8.2).

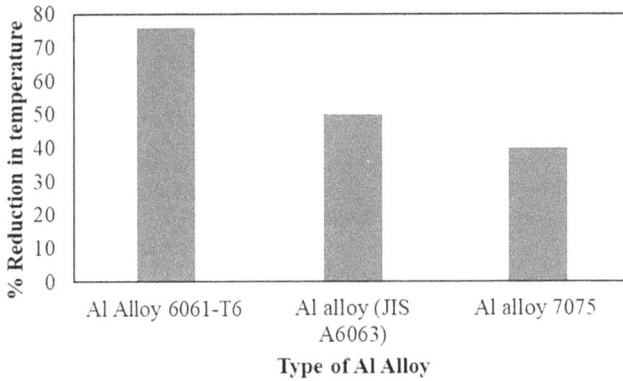

FIGURE 8.2 Percentage reduction in cutting temperature with MQL compared to dry cutting while machining different Al alloys.

8.1.3 EFFECT OF MQL ON TOOL WEAR AND TOOL LIFE

The cutting force and cutting temperature developed during the machining process plays a major role on the tool wear. Due to high temperatures, the tool becomes weak and gradually wears, which leads to tool breakage. Flood cooling effectively removes heat from the machining zone, but keeping in view the adverse effects, MQL is a more preferable cooling technique in machining (Wang et al., 2014). MQL lubrication reduces abrasion, adhesion, and attrition by decreasing the heat due to frictional force at the tool-work interface. The higher MQL flow rate yields marginal adhesion and edge integrity. MQL reduces tool wear and damage at the tool tip, and in turn improves the tool life and productivity. The tool-chip interface temperature is greatly reduced in MQL, thereby protecting the tool from thermal damage. The challenge in MQL machining is difficulty of the lubricant reaching the cutting zone (Yigit, 2014). For example, machining of Al6082 alloy with rapeseed oil in MQL reduced tool wear and enhanced the tool life. Rapeseed oil gives less lubrication while used in MQL during light working loads. Water-based TiO_2 nanofluid used in MQL for machining of aluminum alloy (AA6061-T6) reduced tool wear by lowering the abrasion, adhesion, and attrition. Nanofluids with a small fraction of nanoparticles leads to the formation of a built-up edge due to insufficient lubrication. Nanofluids with a higher fraction of nanoparticles also provide insufficient lubrication due to high density and pumping at a constant pressure. Hence, around 2–3% nanoparticles in the base fluid are maintained based on the nanoparticle and base fluid combination to protect the tool from damage (Najiha et al., 2016) (Table 8.1).

8.1.4 EFFECT OF MQL ON SURFACE ROUGHNESS

In machining of aluminum alloy, BUE formation takes place at lower cutting speeds. This BUE on the one hand favors the formation of compressive residual stress, but on the other side damages the workpiece surface finish. The rectification for damage of surface quality can be done by applying coolant/lubricant and

TABLE 8.1

Benefits of MQL applied to Al alloys

Alloy type	Coolant/lubricant	Remarks
AA6061-T6	Water-based TiO_2 nanofluid	• Effective in decreasing the temperature generated at tool-work interface during machining by producing chilling effect. • Acted as good lubricant and thereby reduced coefficient of friction.
Al 6082	Rapeseed oil	• Interface temperature greatly reduced. • Produced effective lubrication in light loads.

reducing the cutting temperature. In general, surface roughness is less in Al alloys due to high thermal conductivity of the material. As the thermal conductivity of the material is high, the heat is well conducted through the material and the sooner the heat was removed from the cutting zone and dissipated into the ambient environment. Thus, reduction in heat at the cutting zone favors in retaining tool geometry and thus improving the surface roughness (Syahmi et al., 2016). In machining of Al alloys, spindle speed and feed rate are the most influential parameters affecting surface roughness. Increase in spindle speed increases the friction and leads to an increase in temperature. The increase in temperature raises the tool wear and affects the surface roughness (Okonkwo et al., 2016). Higher feed rate increases the surface roughness due to frictional heat. However, with the application of MQL, surface roughness is less due to reduction in frictional heat. The reduction in heat avoids tool blunting and thereby improves the surface finish. For example, CNC milling of Al 6061 with 10% boric acid mixed with SAE 40 base oil in MQL achieved a 20% reduction in surface roughness compared to dry machining. Micro-milling of Al 1100 with the MQL environment using Accu-Lube LB-6000 as a lubricant showed a reduction of 42.85% in the surface roughness (Figure 8.3).

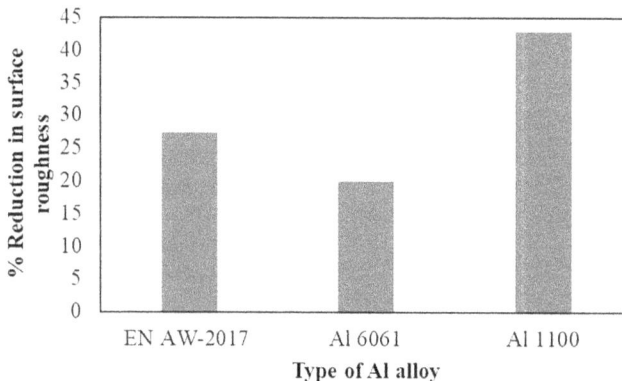

FIGURE 8.3 Reduction percentage in surface roughness with MQL while machining different Al alloys.

8.1.5   Effect of MQL on chip formation

The use of cutting fluids and lubricants greatly affects the chip shape and segmentation. Chip breakability is a major problem in Al alloy machining. Longer chips are not favorable as they damage the machined surface, are a threat to the tool, and a problem to the evacuation system. In machining of Al alloys, the average chip thickness reduces with the rapeseed oil as MQL. The adverse effects like burning of chips and chip segmentation are also reduced due to reduced frictional force between the chip and tool in MQL machining. Chip segmentation occurs randomly and seems to be uniform with a clear shear line between segments in MQL machining (Ekinovic & Ekinovic, 2013). The feed rate, cutting speed, and MQL flow rate are influencing parameters in chip breakability of Al alloy as this is ductile material. For example, MQL machining using Mecagreen 550 lubricant mixed with water as a lubricant enhanced the chip breakability. In the turning of 7075-T6 Al alloy, different chip forms and length are observed in different lubrication conditions. The high temperature chips, when in contact with a coolant, immediately get cooled and this results in strain hardening. This makes the chips brittle and causes breaks. Thus, segmented chips are formed in 7075-T6 Al alloy (Kouam et al., 2015). However, MQL improves overall properties compared with other cooling techniques, but it is not significantly effective in chip removal (Mehner et al., 2022) (Table 8.2).

TABLE 8.2
Effect of MQL on chip formation in Al alloys

Alloy type	Coolant/lubricant	Remarks
Aluminum bronze metal (Cu85.5Al10Fe2.5Mn2)	Vegetable rapeseed oil	• Uniform segmentation of chips • Size of the chip reduced
Al 7075-T6	Mecagreen 550 lubricant (vegetable-based water-soluble oil)	• Feed rate, cutting speed, and MQL flow rate affects the chip formation. • Mecagreen 550 is effective in forming discontinuous chips.

8.2   STEELS

Steel is one of the most ancient and influential materials in human history. Machining of steels is one of the challenging tasks and is addressed by various techniques like conventional cooling, flood cooling, jet-based cooling, cryogenic cooling, air cooling, and minimum quantity lubrication. The current study focuses on the influence of MQL on machinability of steel in different machining processes. MQL is applied in machining of low-carbon steels, medium-carbon steels, and high-carbon steels such as AISI D2 steel, AISI 316 steel, AISI 9310 steel, AISI 1040 steel, and AISI 1045 steel, etc. These materials are machined by different machining processes like turning, milling, grinding, etc. with the use of different types of lubricants or coolants in MQL. Ester oil, Accu lube LB-6000l, Mobil Cut-102 oil, vegetable-based cutting fluids, and nanofluids are applied at different flow rates in MQL.

8.2.1 EFFECT OF MQL ON CUTTING FORCE

In machining of steels, control of cutting force is influenced by machining environment, parameters, and lubrication condition. Cutting forces are influenced by coefficient of friction in machining and can be reduced by minimizing the friction in chip tool interface. Cutting forces are reduced in machining of different steels with different cutting fluids in MQL as the lubricant penetrates into chip tool interface effectively. The penetrated lubricant in the chip-tool interface helps to reduce the coefficient of friction and also causes for embrittlement effect (Masoudi et al., 2018). Reduced cutting temperatures are possible due to effective heat transfer with MQL, which helps to retain the cutting-edge sharpness for a longer time and making the cutting action easier with low cutting force (Dhar et al., 2007). In turning AISI 1040 steel with Mobil Cut-102 oil at an air pressure of 8.0 bar, and lubricant flow rate of 200 ml/h, the cutting force was reduced by 5–15% compared to dry machining. Similarly, when low-carbon steel St52-3 and AISI 1045 steel were machined with ester oil and the MQL coolant, the cutting force was decreased by 17% and 42%, respectively. But when compared with a wet lubricant, the cutting force was decreased by 18%, as shown in Figure 8.4 for AISI 1045.

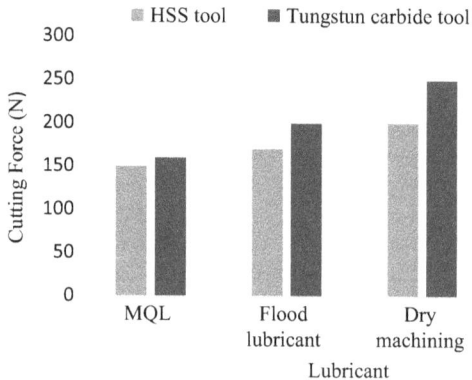

FIGURE 8.4   Variation of cutting force with dry, flood, and MQL lubricant in machining of AISI 1045 steel (Masoudi et al., 2018).

8.2.2 EFFECT OF MQL ON CUTTING TEMPERATURE

A reduction in cutting temperature is observed in MQL compared to dry and wet conditions. This is possible as the penetration of the lubricant is effective with MQL and reduced the coefficient of friction between chip tool interface. The

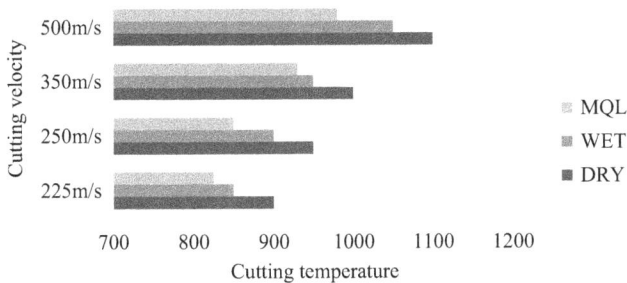

FIGURE 8.5   Variation of cutting temperature with cutting velocity in turning of AISI 9310 steel (d = 1 mm, f = 0.1 mm/rev) (Khan et al., 2009).

effective coefficient of friction between mating surfaces is influenced by the viscosity of the lubricant used in MQL. This reduced friction helps to lower the rubbing action and decreases the heat generation, which leads to lower cutting temperature (Sharma & Sidhu, 2014). While machining at a lower cutting speed, the chip tool interface makes a partial elastic contact zone, making the MQL penetrate easily due to capillary action, but as speed increases, the chip makes full bulk or plastic contact and decreases the rate of MQL penetration in the chip tool interface (Dhar et al., 2007). Turning AISI 1040 steel with Mobil Cut-102 oil-based MQL decreases the cutting temperature by 5–15% than dry machining, depending on the process and cutting parameters (Dhar et al., 2007). When turning AISI 9310 alloy steel with MQL-based lubricant with food-grade vegetable oil shows an average cutting temperature reduction of 5–10% compared to wet lubrication (Khan et al., 2009). The rate of MQL cooling is influenced by coolant flow rate, coolant pressure, and standoff distance. With the increase in coolant pressure, there is a reduction in cutting temperature, and the maximum reduction was observed at 5 bars at a cutting speed of 150 m/min (Yassin & Teo, 2015). The temperature was reduced by 5–12%, with 15–55 mm standoff distance and temperature reduction were observed to be a maximum at a 55 mm standoff distance (Kumar & Singh, 2013). Thus, MQL reduces the cutting zone temperature that eventually decreases flank wear and built-up edge formation (Figure 8.5).

8.2.3   EFFECT OF MQL ON TOOL WEAR AND TOOL LIFE

The function of cutting fluid is to act as a lubricant when introduced in the machining zone, and then reducing the coefficient of friction and thus reducing the cutting force and cutting temperature. Reduction in cutting zone temperature in a chip tool interface causes the tool to retain hardness for a long time, decreasing the abrasive wear of the cutting tool (Dhar et al., 2007). Flank wear takes place as feed and speed increase irrespective of cutting fluid, but due to more penetration of the MQL lubricant in the chip tool interface forced convection and lubrication to occur and thus reduced the flank wear. This reduction

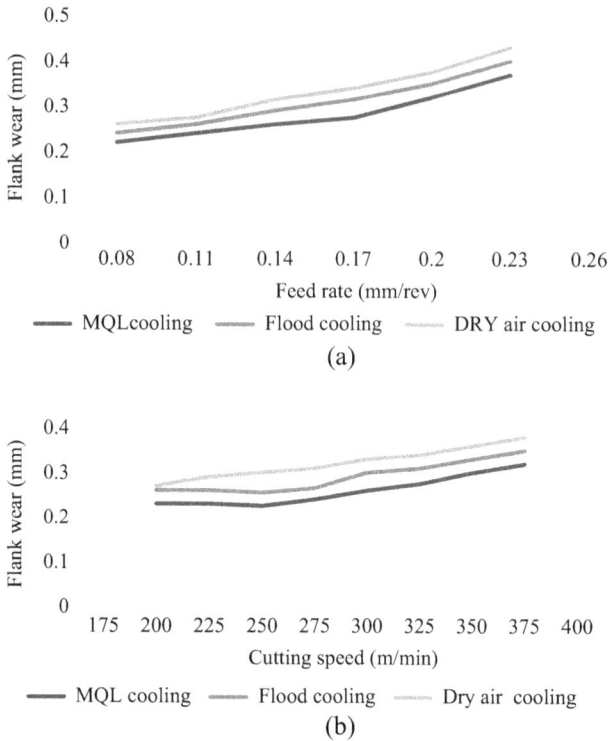

FIGURE 8.6 Variation of flank wear with (a) feed rate and (b) cutting speed (Sankar & Choudhury, 2015).

in flank wear will help to increase the tool life. Introduction of MQL in ma-chining operations reduces the tool wear and improves the life compared to dry and wet conditions. On an average, there is a 15% and 6% reduction in flank wear when compared to dry and wet lubrication, respectively, at a similar feed rate. Similarly, a 16% and 11% reduction of flank wear was observed when compared with dry and wet lubricant, respectively, at a similar cutting speed, causing the tool life to increase (Sankar & Choudhury, 2015). Increased cooling pressure of the coolant in MQL decreases the tool wear (Naves et al., 2013). While turning ASI 1045 mild steel under MQL with varying pressure from 1–5 Mpa and nozzle angle of 5°–50°, the reduction of tool wear is 35% at 5 Mpa and 50° (Yassin & Teo, 2015) (Figure 8.6).

8.2.4 Effect of MQL on surface roughness

Surface roughness is mostly influenced by the machining process and tool-workpiece interaction. Surface roughness is less in MQL due to the effective penetration of lubricant at the chip tool interface. Reduced tool wear and tool tip damage in MQL is also helpful in lower surface roughness compared to dry and wet

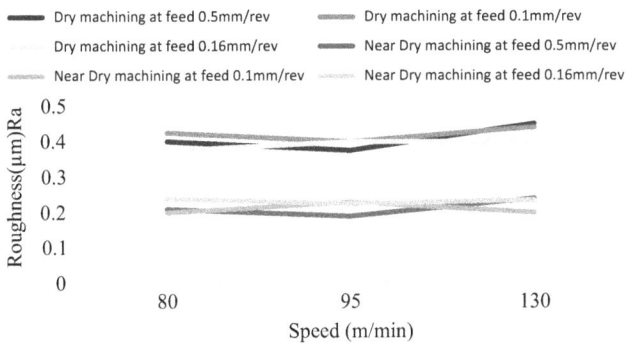

FIGURE 8.7   Effect of near dry and dry machining on surface roughness at different feeds and speeds (Sharma & Sidhu, 2014).

machining (Khan et al., 2009). While grinding hardened steel, clogging action in the wheel takes place due to ductile property, thus increasing roughness while machining with MQL. However, the required quantity of coolant is less with MQL (Lopes et al., 2019). Surface roughness is reduced by 23.5% and 14% than dry and flood lubricants in turning of micro-alloy steel with emulsion-type mineral MQL oil (Sankar & Choudhury, 2015). It is noted that the feed rate has the most influence on surface roughness and desired results are obtained at a 0.5 mm/rev feed rate when AISI D2 steel was machined with Accu lube LB-6000l MQL oil (Sharma & Sidhu, 2014) (Figure 8.7).

8.2.5   Effect of MQL on chip morphology

MQL helps to reduce coefficient of friction between chip-tool interface and decreases the heat generation, which may lead to reduction in chip-tool interface temperature and change the behavior of chip formation. As the secondary heat generation zone i.e., chip tool interface plays a very important role in deciding the shape and behavior of chip formation and MQL influences the chip formation process significantly (Khan et al., 2009). In a dry machining condition, formation of a built-up edge and sticking of chip with the tool may cause an increase in chip thickness, whereas in wet machining due to ineffective penetration of cutting fluid at a high cutting speed increases the chip thickness. The chips produced under both dry and wet conditions are of half turn and helical, respectively, due to high cutting temperature and chip tool interactions. However, MQL prevents the formation of a built-up edge and also reduces chip thickness by penetrating the chip tool interface effectively and forms a ribbon-type chip, as shown in Figure 8.8. In turning AISI 9310 steel using food-grade vegetable oil cutting fluid with MQL, ribbon-type continuous chips at lower feed rates and more or less tubular type continuous chips at higher feed rates are observed. In turning micro-alloyed steel under MQL with emulsion-type mineral oil, chip thickness is reduced by 4.5% and 3.0%, respectively, compared to dry and wet machining due to effective penetration (Sankar & Choudhury, 2015).

S.NO	Chip shape	Group	Machining condition
1		Half turn	Dry
2	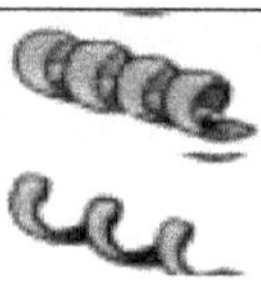	Tabular or Helical	Wet
3		Ribbon	MQL

FIGURE 8.8 Comparison of chip shape under dry, wet, and MQL by vegetable oil conditions (Khan et al., 2009).

8.3 TITANIUM

Titanium alloy is most widely used material in aerospace and automobile industries due to its high strength-to-weight ratio and corrosive resistance. While machining the titanium alloy, the temperatures are high due to low thermal conductivity of the titanium alloys and it reduces the tool life. Due to high cutting temperatures, the surface quality also deteriorates. Hence, MQL is used to improve the machining performance in machining titanium alloy. Various cutting fluids, such as vegetable oil, emulsifiers, palm oil, and synthetic esters are used in MQL while titanium machining. The effect of MQL on the performance of titanium machining is discussed in this section.

8.3.1 EFFECT OF MQL ON CUTTING FORCES

Cutting force plays a significant role in machining to shear the material and is greatly affected by the cutting environment, feed, depth of cut, and the cutting speed. In MQL, fluid parameters like nozzle diameter, elevation angle, and mass flow rate also show an effect on the cutting force (Zan et al., 2021). Cutting forces are reduced with the application of MQL in turning, milling, drilling, and grinding of titanium alloy components. With the application of MQL, the lubrication effect is better due to the fluid droplets that travel with high speed with compressed air and ease of flow of cutting fluid into the tool workpiece interface. This helps to reduce the coefficient of friction and decreases the cutting force (Mishra et al., 2020). This is because the

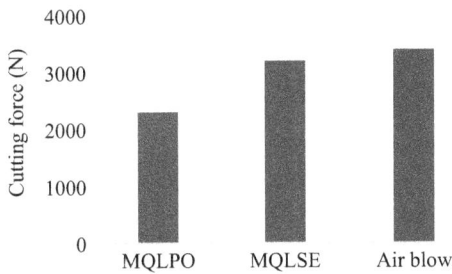

FIGURE 8.9   Cutting forces with various coolant/lubricant conditions (Rahim & Sasahara, 2011).

lubricant helps to reduce heat generated during cutting, which can lead to lower tool wear and reduced cutting forces. For example, palm oil in MQL while drilling of titanium alloy reduced effectively compared to synthetic ester and air blow conditions (Figure 8.9). The cutting forces are 30% and 6.5% less with palm oil and synthetic ester, respectively, compared to air blow condition (Rahim & Sasahara, 2011).

8.3.2   EFFECT OF MQL ON CUTTING TEMPERATURES

Cutting temperature is influenced by the amount of heat generated at the chip-tool interface. In machining of difficult-to-cut materials like titanium alloy, cutting temperatures play a vital role. The cutting temperatures are high in the machining of titanium alloy because of its low thermal conductivity. In general, dry machining shows high cutting temperatures due to a high coefficient of friction and absence of cutting fluids. By reducing the coefficient of friction between the chip tool interface with MQL, the amount of heat generation can be reduced in titanium machining. Also, with pressurized flow of fluid in MQL, heat is carried out by the cutting fluid from the tool work interface with MQL in titanium machining. Hence, the cutting temperature is less in titanium machining with MQL compared to dry and conventional machining. At high cutting speeds and feeds, the temperatures are more compared to low speeds and feeds, irrespective of cutting fluid conditions (Rahim & Sasahara, 2011). The cutting temperatures are reduced by 66% in MQL compared to dry conditions (Sahoo et al., 2021). By applying water as MQL, the cutting temperatures are decreased at different cutting speeds compared to dry and pressurized air supply (Figure 8.10).

8.3.3   EFFECT OF MQL ON SURFACE ROUGHNESS

Surface roughness depends on the tool tip temperature and the tool wear. High temperatures lead to excessive tool wear and to high surface roughness. By using MQL, the cutting temperature reduces and controls the tool wear, which helps to improve the surface quality of the component. Surface roughness is reduced with MQL application in turning, milling, grinding, and drilling of titanium alloy components. Even at a high speed and feed, the surface roughness is reduced with the

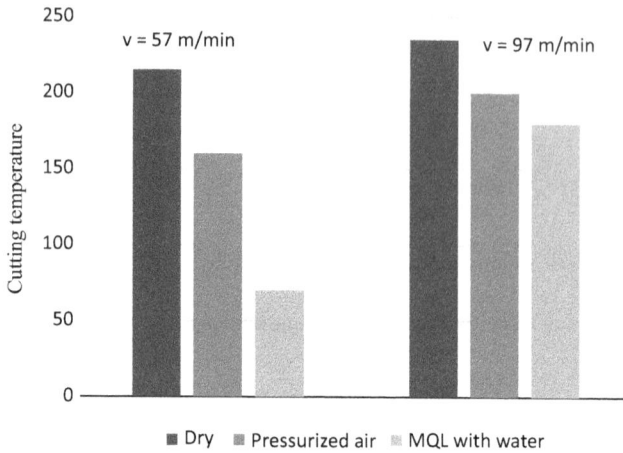

FIGURE 8.10   Tool tip temperatures at different machining conditions (Sahoo et al., 2021).

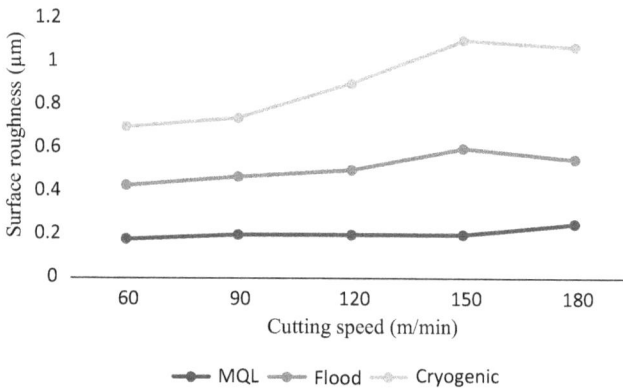

FIGURE 8.11   Surface roughness at different cutting environments (Shokrani et al., 2019).

application of MQL (Deiab et al., 2014). Surface roughness of milled titanium alloy reduced under the MQL with vegetable oil at a high speed and feed compared with a dry machining condition (Khaliq et al., 2020). Figure 8.11 represents the surface roughness of titanium alloy at different conditions with respect to cutting speed with the application of MQL (Shokrani et al., 2019). The use of synthetic oils in MQL gives a cooling effect in the machining and is most suitable while machining titanium alloy to decrease the surface roughness on the component (Sadeghi et al., 2009).

8.3.4   EFFECT OF MQL ON TOOL WEAR AND TOOL LIFE

Tool wear is more in machining hard materials like titanium alloys as it occurs mainly because of the high cutting temperatures generation. These high cutting temperatures in titanium machining are due to low thermal conductivity of the

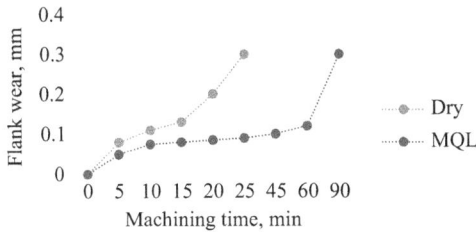

FIGURE 8.12 Flank wear with respect to machining time (Swain et al., 2022).

workpiece material, which leads to more accumulated heat in the tool. This accumulated heat smoothens the tool material and leads to wear on the tool faces. With an increase of tool wear, surface roughness of the component increases as well as power consumption. As the cutting forces and cutting temperatures are reduced with the application of MQL, it helps to reduce the tool wear compared to dry machining (Deiab et al., 2014). Tool wear also depends on the cutting speed as cutting temperatures are increased with cutting speed. However, with the application of water in MQL, even at higher cutting speeds the tool wear is less compared to dry machining (Swain et al., 2022). Tool wear is reduced with a coated tool or uncoated tool by the usage of MQL (Liu et al., 2013). Similarly, tool life is more with MQL, as it is influenced by the tool wear in machining (Figure 8.12). Abrasion wear is dominant on the rake and flank face of the tool, which is low with MQL compared to dry machining. Chip adhesion and nose wear are also dominant wear mechanisms that were less with MQL compared to other conditions. By considering all these wear mechanisms, the MQL machining condition shows the least wear and is more suitable for productivity of titanium alloy components (Khatri & Jahan, 2018).

8.3.5 EFFECT OF MQL ON CHIP MORPHOLOGY

In titanium machining, chip formation attains importance due to low modulus of elasticity of the workpiece material. Dry machining of titanium alloys leads to formation of ribbon-like chips due to high temperature accumulation at the cutting zone. Pressurized air supply helps to form irregular types of helical chips, whereas water MQL causes chips to deflect in helical shapes without accumulating at the tool tip, due to effective cutting heat dissipation (Sahoo et al., 2021). Effective chip thickness and shear area are small in MQL because of improved lubrication and cooling actions, (Swain et al., 2022). Chips formed in dry machining, with pressurized air supply and water MQL, are presented in Figure 8.13.

8.4 HARD MATERIALS

MQL and NF-MQL are used in machining of hard-to-cut materials and their alloys, namely GH 4099, Inconel X-750, GH 4169, etc. to improve the machining process. These materials are machined with processes like milling, turning, and grinding using vegetable oils, oil + water, and nanoparticles as MQL fluids.

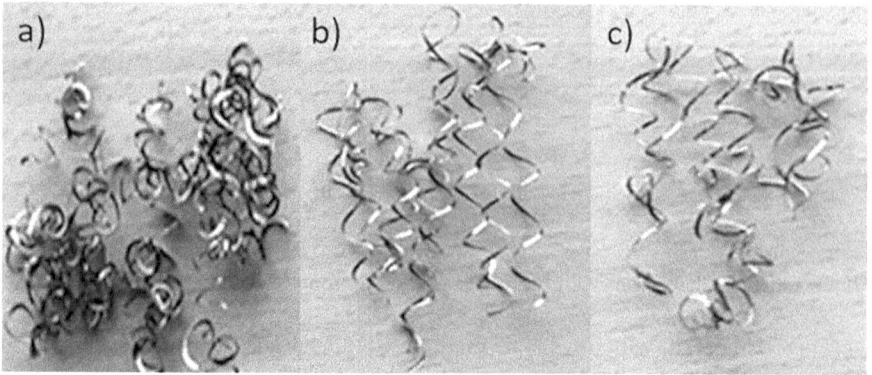

FIGURE 8.13 Chips formed in different cutting conditions: a) dry machining, b) pressurized air supply, and c) water MQL (Sahoo et al., 2021).

8.4.1 EFFECT OF MQL ON CUTTING FORCE

With the application of vegetable oil, a thin film formation occurs at the tool chip interface, which reduces the coefficient of friction. When these oils are mixed with water, due to the expandability of the oil molecules, thousands of them are arranged together, forming a thin oil film on the water molecule as small water droplets. These are sprayed at the cutting area with a certain speed to cool the cutting zone and remove chips. Cutting forces in the milling of GH4099 are reduced by about 12.8% and 28.2% with vegetable oil and vegetable oil mixed with water, respectively (Sun et al., 2022). Figure 8.14 shows the variation of cutting force reduction in milling using the MQL technique compared to dry cutting. Milling of superalloy X-750 using a base fluid (Belgin oil) and base fluid mixed with nanoparticles (hBN)

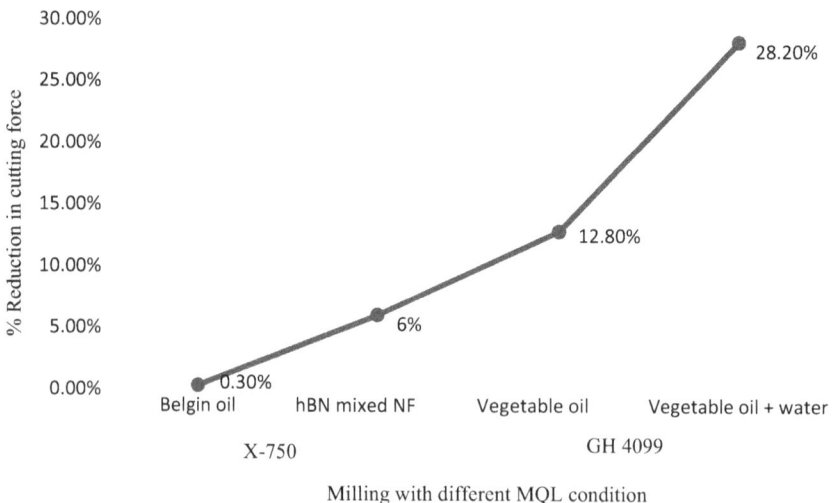

FIGURE 8.14 Variation of cutting force in milling under different MQL condition.

reduced the cutting force by 0.3% and 6%, respectively, compared to dry milling (Şirin et al., 2021). The performance of hBN was better due to its high thermal conductivity and chemical inertness as well. Using different vegetable oils as base fluids in grinding found that tangential grinding force and normal grinding force were reduced 38.88% and 31.16%, respectively, relative to dry grinding (Li et al., 2016).

8.4.2 EFFECT OF MQL ON CUTTING TEMPERATURE

The use of vegetable oil as a base fluid in MQL reduces the heat exchange capability due to high viscosity, thus resulting in a higher cutting temperature. However, cutting temperature is less compared to the dry machining temperature. MQL can reduce the cutting temperature to a better extent in hard-to-cut material compared to dry cutting. But, an advanced MQL technique like cryoMQL gives a better performance and efficient cooling and lubrication. MQL, cryo, and cryo MQL (cryogenic cooling + MQL) techniques are used in machining of alloys like Inconel 625, Inconel 718, and GH 4169. These materials were machined with liquid nitrogen in cryo and cryo MQL for turning while vegetable oil is used as base fluid in MQL for grinding. Cutting temperatures that occurred in cryo cooling and CryoMQL are reduced by 21.7% and 24.9%, respectively, compared to the MQL (water-soluble cutting oil formulated with vegetable esters and special additives were employed during the machining) (Yıldırım et al., 2020). Figure 8.15 shows the variation of grinding temperature in (°C) for seven different kinds of vegetable oils used.

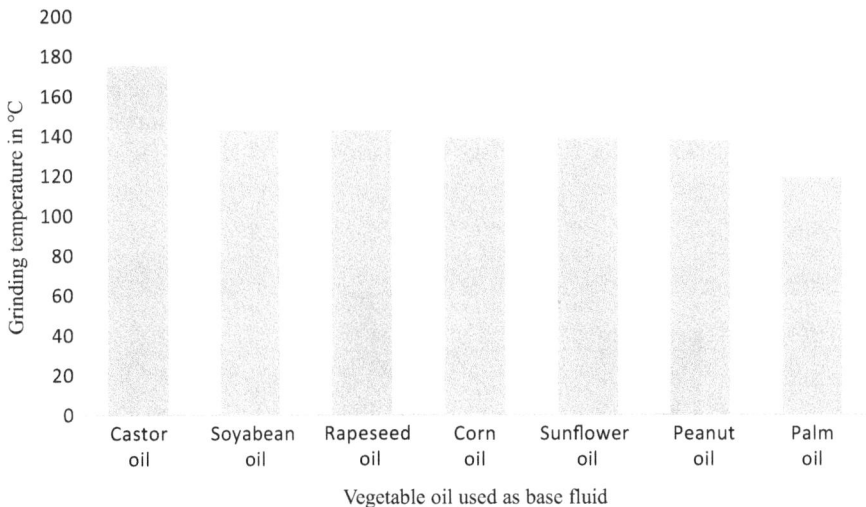

FIGURE 8.15   Grinding temperature variation with different vegetable oils (Li et al., 2016) (at peripheral speed of grinding wheel Vs = 30 m/s, feed speed Vw = 3,000 mm/min, cutting depth ap = 10 μm, MQL flow rate = 50 mL/h, MQL nozzle distance = 12 mm, MQL nozzle angle = 15°, MQL gas pressure = 0.6 MPa).

8.4.3 EFFECT OF MQL ON SURFACE ROUGHNESS

Various machining operations like turning, milling, and grinding were performed with MQL, NF-MQL, cryoMQL, and EMQL (electrostatic MQL) on materials like Inconel 625, Inconel 718, X-750, and GH4169 for improving the surface finish. In machining of these materials, liquid nitrogen is used in cryogenic cooling, cryoMQL and hBN are used as a nano additive in nanofluid MQL. When soyabean oil was mixed with MWCNTs in machining of Inconel 625, the surface finish was improved compared to dry cutting due to an increase in cooling and heat transfer rate from the tool-workpiece-chip interface because of their high thermal conductivity. The surface finish of MWCNT mixed in soyabean oil (nanofluid) was improved by 55.58% and 5.48%, respectively, compared to dry cutting and conventional flood machining (Hegab et al., 2018). The MQL method in machining operations, especially with carbide cutting tools, plays an effective role in reducing friction and wear by creating a very thin layer on the tool-chip contact regions and thus improves the surface finish. hBN mixed nanofluid is found to improve the surface quality by the addition of hBN nanoparticles as they carry the load and thus avoid the contact of friction pairs. Surface roughness was reduced by 39% and 47% with the base fluid (Belgin oil) and hBN mixed nanofluid (Şirin et al., 2021). Figure 8.16 shows the reduction in surface roughness under the application of different MQL cutting conditions in different processes. MQL with vegetable oils has a better surface finish compared to flood cooling because the formation of furrows on the workpiece are reduced and narrowed significantly in MQL. Compared to cryogenic machining, Ra values obtained by MQL and CryoMQL decreased by 13.8% and 24.82%, respectively (Yıldırım et al., 2020). Grinding of GH4169 (Ni-based alloy) using seven different vegetable oils in MQL exhibited better surface finish compared to flood cooling because of the same reason as explained above. In comparison to other oils, castor oil gives the best result. In EMQL, increasing the cutting speed from 60 m/min to 120 m/min causes the surface roughness to reduce by 8%. The best surface roughness Ra is achieved under EMQL conditions at a 120 m/min cutting speed due to the electron mirror effect that enhances lubrication on the tool surface (De Bartolomeis et al., 2021).

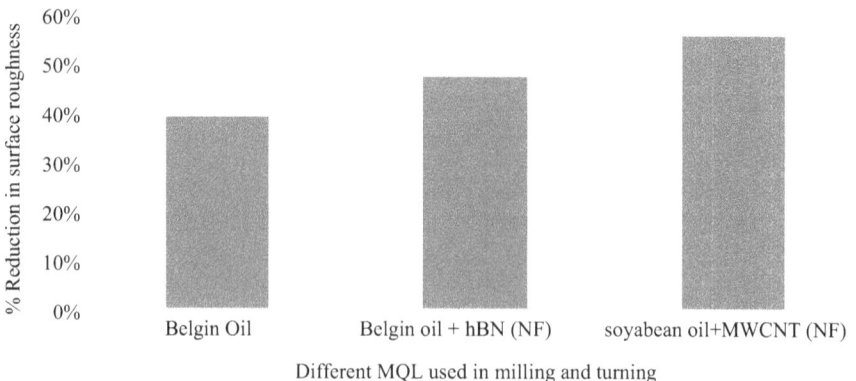

Different MQL used in milling and turning

FIGURE 8.16   The percentage reduction in surface roughness under different MQL conditions.

8.4.4 Effect of MQL on tool wear and tool life

In dry machining of hard-to-cut material due to the high heat generation at the tool-chip interface, the tool wears more rapidly and hence the tool life reduces. With the use of different MQL techniques, such as base fluid MQL, cryoMQL, nanofluid MQL, and MQCL (minimum quantity cooling lubrication), such problems can be resolved. Different machining operations like turning and milling are performed on various hard-to-cut materials such as Inconel 625 and Inconel 718. Water-soluble cutting oil formulated with vegetable esters and special additives used in MQL wraps the cutting region with a layer of oil and this helps to reduce friction and contributes to the heat transfer due to the evaporation of droplets. The combination of both higher penetration of oil and higher heat transfer of very low-temperature oil increases the efficiency of MQL. Tool wear is reduced by 18.17% and 4.54% under a MWCNT mixed nanofluid in the turning of Inconel 625 compared to dry machining and wet cutting conditions, respectively (Yıldırım et al., 2020). Tool life under the MQCL cutting condition is 1.57 times more than the dry cutting condition (Zhang et al., 2012). When the MQCL jets hit the tool and workpiece surface at high speed and high pressure, the cutting oil can be fragmented into micro-droplets and this leads to penetration of the cutting oil at the tool/chip and tool/workpiece interfaces. Fragmented oils have lower shear strength than the chip material, thus reducing friction. At the same time, the cryogenic compressed air is sufficient to cool the workpiece material and the cutting edges, though it may not be able to flush away the chips. Tool wear (VB = 0.211 mm) is observed to be at a minimum in cryoMQL at a cutting speed of 75 m/min. Moreover, tool wear is decreased by 50.67% and 79.60% with MQL and cryoMQL compared with cryogenic machining (Hegab et al., 2018).

8.4.5 Effect of MQL on chip morphology

Chip morphology provides important clues about the cutting mechanics and is closely related to surface integrity of the finished product (surface roughness, surface topography, etc.) and machining efficiency. Turning of hard-to-cut material such as Inconel 625 and Inconel 718 under base fluid, base fluid mixed with nanoadditives of hBN, MWCNT, and Al_2O_3, cryogenic cooling, and cryoMQL were carried out. During dry machining, short helical and burnt blue color chips were formed, while in wet machining long helical and light blue color continuous chips were formed, and using nanofluid segmented helical chips of gray color were formed due to a reduction of cutting zone temperature and high velocity jet (Hegab et al., 2018). MWCNT's nanofluid provides larger chip helix angles compared to the Al_2O_3 nanofluid and dry turning due to formation of a hydrodynamic layer between the chip and tool rake face. Long helical chips in dry, more helix angle chips in Al_2O_3 and discontinuous chips in MWCNTs were formed. BUE was found on the chip in dry machining due to ineffective cooling and lubrication, while no BUE was formed when machining with nanofluids due to improved cooling and lubrication at a cutting speed of 60 m/min and feed rate of 0.3 mm/rev. Figure 8.17 shows the effect of different MQLs in chip morphology. In cryogenic cooling, large scratches were seen on the back face of the chip, owing to the severe friction

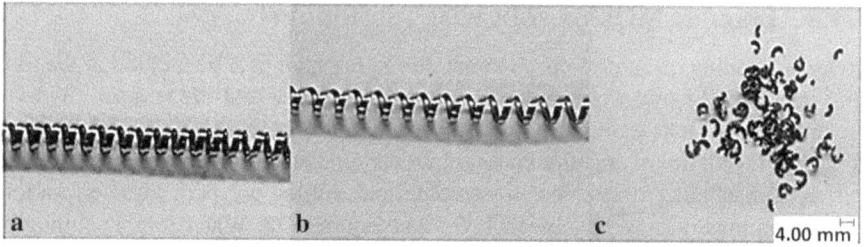

FIGURE 8.17 Chips during machining of Inconel 718: (a) without nanoadditives, (b) Al_2O_3 nanofluid, (c) MWCNT's nanofluid (Hegab et al., 2018) (cutting speed of 60 m/min and feed rate of 0.3 mm/rev).

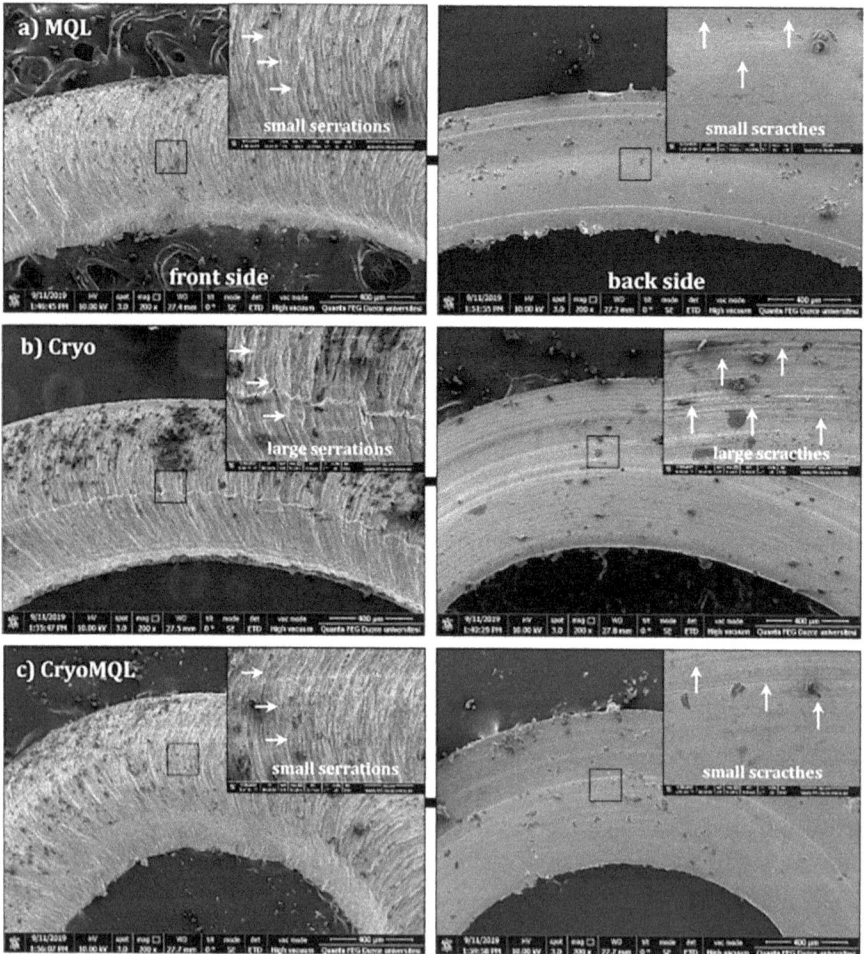

FIGURE 8.18 Chip morphologies of front and back side when using (a) MQL, (b) cryo, and c) cryo MQL (Yıldırım et al., 2020) (cutting speed of 75 m/min and feed rate of 0.1 mm/rev).

between the tool rake surface and the chip, which shows insufficient lubrication. Water-soluble cutting oil formulated with vegetable esters and special additives (MQL) and hybrid cooling that is cryoMQL significantly reduces the scratches on the back face of the chip due to reduction in friction and superior lubrication. On the front face of the chip, large, linear, and sharp serrations were found in cryogenic cooling compared to MQL and cryoMQL (Yıldırım et al., 2020). Figure 8.18 shows the effect of different MQLs on the chip morphology. The front and back sides of chips were examined at a cutting speed of 75 m/min.

REFERENCES

De Bartolomeis, A., Newman, S. T., & Shokrani, A. (2021). High-speed milling Inconel 718 using Electrostatic Minimum Quantity Lubrication (EMQL). *Procedia CIRP*, 101, 354–357.

Deiab, I., Raza, S. W., & Pervaiz, S. (2014). Analysis of lubrication strategies for sustainable machining during turning of titanium Ti-6Al-4V alloy. *Procedia CIRP*, 17, 766–771.

Dhar, N. R., Ahmed, M. T., & Islam, S. (2007). An experimental investigation on effect of minimum quantity lubrication in machining AISI 1040 steel. *International Journal of Machine Tools and Manufacture*, 47(5), 748–753.

Ekinović, Sabahudin, et al. "Cutting forces and chip shape in MQL machining of Aluminium Bronze." space 6 (2013): 6.

Hegab, H., Umer, U., Soliman, M., & Kishawy, H. A. (2018). Effects of nano-cutting fluids on tool performance and chip morphology during machining Inconel 718. *The International Journal of Advanced Manufacturing Technology*, 96. 10.1007/s00170-018-1825-0.

Junge, T., Mehner, T., Nestler, A., Schubert, A., & Lampke, T. (2022). Surface properties in turning of aluminum alloys applying different cooling strategies. *Procedia CIRP*, 108(C), 246–251, doi: 10.1016/j.procir.2022.03.043.

Khaliq, W., et al. (2020). Tool wear, surface quality, and residual stresses analysis of micro-machined additive manufactured Ti–6Al–4V under dry and MQL conditions. *Tribology International*, 151, 106408.

Khan, M. M. A., Mithu, M. A. H., & Dhar, N. R.(2009). Effects of minimum quantity lubrication on turning AISI 9310 alloy steel using vegetable oil-based cutting fluid. *Journal of materials processing Technology*, 209(15–16), 5573–5583.

Khatri, A., & Jahan, M. P. (2018). Investigating tool wear mechanisms in machining of Ti-6Al-4V in flood coolant, dry and MQL conditions. *Procedia Manufacturing*, 26, 434–445.

Khettabi, R., Nouioua, M., Djebara, A., & Songmene, V. (2017). Effect of MQL and dry processes on the particle emission and part quality during milling of aluminum alloys. *Int. J. Adv. Manuf. Technol.*, v92(5–8), 2593–2598, doi: 10.1007/s00170-017-0339-5.

Kilickap, E., Huseyinoglu, M., & Ozel, C. (2011). Empirical Study Regarding the Effects of Minimum Quantity Lubricant Utilization on Performance Characteristics in the Drilling of Al. XXXIII(1), 52–57.

Kouam, J., Songmene, V., Balazinski, M., & Hendrick, P. (2015). Effects of minimum quantity lubricating (MQL) conditions on machining of 7075-T6 aluminum alloy. 1325–1334, doi: 10.1007/s00170-015-6940-6.

Kumar, A., Singh, G., & Gill, S. S. (2013). Impact of varying the nozzle stand-off distance on cutting temperature in turning of EN-31 steel with minimum quantity lubrication. *Proc IJRET*, 2(6), 931–936.

Li, B., Li, C., Zhang, Y., Wang, Y., Jia, D., & Yang, M. (2016). Grinding temperature and energy ratio coefficient in MQL grinding of high-temperature nickel-base alloy by using different vegetable oils as base oil. *Chinese Journal of Aeronautics*, 29(4), 1084–1095.

Liu, Z., An, Q., Xu, J., Chen, M., & Han, S. (2013). Wear performance of (nc-AlTiN)/(a-Si3N4) coating and (nc-AlCrN)/(a-Si3N4) coating in high-speed machining of titanium alloys under dry and minimum quantity lubrication (MQL) conditions. *Wear*, 305(1-2), 249–259.

Liu, Z. Q., Cai, X. J., Chen, M., & An, Q. L. (2010). Investigation of cutting force and temperature of end-milling Ti – 6Al – 4V with different minimum quantity lubrication (MQL) parameters. 225, 1273–1279, doi: 10.1177/2041297510393793.

Lopes, J. C., et al. (2019). Behavior of hardened steel grinding using MQL under cold air and MQL CBN wheel cleaning. *The International Journal of Advanced Manufacturing Technology*, 105(10), 4373–4387.

Masoudi, S., et al. (2018). Experimental investigation into the effects of nozzle position, workpiece hardness, and tool type in MQL turning of AISI 1045 steel. *Materials and Manufacturing Processes*, 33(9), 1011–1019.

Mehner, T., Stief, P., Dantan, J., Etienne, A., & Siadat, A. (2022). ScienceDirect ScienceDirect Surface properties in turning of aluminum alloys applying different cooling Surface properties turning of aluminum applying May alloys France different cooling strategies strategies new methodology to analyze the functional and. *Procedia CIRP*, 108, 246–251, doi: 10.1016/j.procir.2022.03.04.

Mishra, S. K., Ghosh, S., & Aravindan, S. (2020). Machining performance evaluation of Ti6Al4V alloy with laser textured tools under MQL and nano-MQL environments. *Journal of Manufacturing Processes*, 53, 174–189.

Najiha, M. S., Rahman, M. M., & Kadirgama, K. (2016). Performance of water-based TiO 2 nano fl uid during the minimum quantity lubrication machining of aluminium alloy, AA6061-T6. *J. Clean. Prod.*, 135, 1623–1636, doi: 10.1016/j.jclepro.2015.12.015.

Naves, V. T. G., Da Silva, M. B., & Da Silva, F. J. (2013). Evaluation of the effect of application of cutting fluid at high pressure on tool wear during turning operation of AISI 316 austenitic stainless steel. *Wear*, 302(1-2), 1201–1208.

Okonkwo, U. C., Okokpujie, I. P., Sinebe, J. E., & Ezugwu, C. A. K. (2016). Comparative analysis of aluminium surface roughness in end-milling under dry and minimum quantity lubrication (MQL) conditions. doi: 10.1051/mfreview/2015033.

Rahim, E. A., & Sasahara, H. (2011). A study of the effect of palm oil as MQL lubricant on high speed drilling of titanium alloys. *Tribology International*, 44(3), 309–317.

Sadeghi, M. H., et al. (2009). Minimal quantity lubrication-MQL in grinding of Ti–6Al–4V titanium alloy. *The International Journal of Advanced Manufacturing Technology*, 44(5), 487–500.

Sahoo, S. P., Datta, S., Roy, T., & Ghosh, S. (2021). Machining performance of Ti6Al4V under dry environment, pressurized air supply and water-MQL: analysis of machining-induced vibration signals and captured thermographs. *Sādhanā*, 46(4), 1–22.

Sankar, M. R., & Choudhury, S. K. (2015). Experimental study and modeling of machining with dry compressed air, flood and minimum quantity cutting fluid cooling techniques. *Procedia CIRP*, 31, 228–233.

Sharma, J., & Sidhu, B. S. (2014). Investigation of effects of dry and near dry machining on AISI D2 steel using vegetable oil. *Journal of cleaner production*, 66, 619–623.

Shokrani, A., Al-Samarrai, I., & Newman, S. T. (2019). Hybrid cryogenic MQL for improving tool life in machining of Ti-6Al-4V titanium alloy. *Journal of Manufacturing Processes*, 43, 229–243.

Şirin, Ş., Sarıkaya, M., Yıldırım, Ç. V., & Kıvak, T. (2021). Machinability performance of nickel alloy X-750 with SiAlON ceramic cutting tool under dry, MQL and hBN mixed nanofluid-MQL. *Tribology International*, 153, 106673.

Sun, H., Zou, B., Chen, P., Huang, C., Guo, G., Liu, J., … & Shi, Z. (2022). Effect of MQL condition on cutting performance of high-speed machining of GH4099 with ceramic end mills. *Tribology International*, 167, 107401.

Swain, S., Kumar, R., Panigrahi, I., Sahoo, A. K., & Panda, A. (2022). Machinability performance investigation in CNC turning of Ti-6Al-4V alloy: Dry versus Iron-aluminium oil coupled MQL machining comparison. *International Journal of Lightweight Materials and Manufacture*.

Syahmi, A., Bin, A., Bahar, R., & Ariff, T. F. (2016). Minimum Quantity Lubrication in Micromachining: A Greener Approach. 1782–1787.

Wang, C., Chen, M., An, Q., Wang, M., & Zhu, Y. (2014). Tool Wear Performance in Face Milling Inconel 182 using Minimum Quantity Lubrication with Different Nozzle Positions. 15(3), 557–565, doi: 10.1007/s12541-014-0371-4.

Yassin, A., & Teo, C. Y. (2015). Effect of pressure and nozzle angle of minimal quantity lubrication on cutting temperature and tool wear in turning. *Applied Mechanics and Materials*, 695. Trans Tech Publications Ltd.

Yigit, R. (2014). An experimental investigation of effect of minimum quantity lubrication in machining 6082 aluminum alloy. 29–33.

Yoshimura, H., Itoigawa, F., & Nakamura, T. (2005). Development of Nozzle System for Oil-on-Water Droplet Metalworking Fluid and Its Application to Practical Production Line *. 48(4), 723–729.

Yıldırım, Ç.V., Kıvak, T., Sarıkaya, M., & Şirin, Ş. (2020). Evaluation of tool wear, surface roughness/topography and chip morphology when machining of Ni-based alloy 625 under MQL, cryogenic cooling and CryoMQL. *Journal of materials research and technology*, 9, 2079–2092.

Zan, Z., Guo, K., Sun, J., Wei, X., Tan, Y., & Yang, B. (2021). Investigation of MQL parameters in milling of titanium alloy. *The International Journal of Advanced Manufacturing Technology*, 116(1), 375–388.

Zhang, S., Li, J. F., & Wang, Y. W. (2012). Tool life and cutting forces in end milling Inconel 718 under dry and minimum quantity cooling lubrication cutting conditions. *Journal of cleaner production*, 32, 81–87.

9 MQL with micro and nanofluids

9.1 SOLID LUBRICANT ADDITIVES

Solid additives are in the form of finely dispersed particles, generally used for friction and wear protection. These are capable of decreasing cutting temperature and improving the surface finish by reducing the coefficient of friction between the tool and workpiece. These are also used as additives in micro- and nano-size in different base fluids. Type of solid lubricant, amount of solid lubricant content, and size of the particle in the coolant are the influencing parameters in machining (Zailani et al., 2011). Some examples of solid lubricants are molybdenum disulfide (MoS_2), polytetrafluoroethylene (PTFE), graphite, boric acid (H_3BO_3), boron nitride, tungsten disulfide, cerium fluoride, calcium fluoride (CaF_2), talc, etc. The selection of proper lubricant can improve the machining performance by reducing the friction in the machining zone. Micro- and nanofluids with different solid lubricant additives are used in machining of materials like ADC12 aluminum alloy, Al 7075 alloy, TC4 alloy, EN8 steel, AISI1040 steel, AISI1045 steel, Ti-6Al-4V alloy, etc. Solid lubricants are used as additives in micro- and nano-size in the base oils, such as castor oil, rapeseed oil, corn oil, soybean oil, palm oil, coconut oil, Accu-Lube oil, SAE 40 oil, water-soluble oil, glycol, etc. These micro and nanofluids are used as coolants/lubricants in different machining operations like drilling, tapping, milling, turning, grinding, etc.

9.1.1 EFFECT ON CUTTING FORCE

In the machining operation, cutting forces are significantly reduced due to low coefficient of friction between the chip-tool interface by solid lubricant particles like boric acid, graphite, etc., compared to dry and wet lubricants (Rao & Krishna, 2008). The formation of lubrication layer, unique bond characteristics, and ability to retain structure at higher temperatures are the main reasons to reduce the cutting force significantly (Krishna & Rao, 2008). Boric acid causes more cutting force reduction compared to graphite due to early film formation and smearing action during machining at lower values of machining parameters (Vamsi Krishna et al., 2010). In turning of AISI 1040 steel, boric acid (H_3BO_3) and molybdenum disulfide (MoS_2) micro- and nano-size particles mixed with coconut and sesame oils reduced the cutting forces. Nano boric acid and nano MoS_2 reduced the cutting force by 38% and 43% compared with micro-particle additives (Padmini et al., 2015). Nano MoS_2 in base oil reported less cutting force due to Brownian action and improved heat exchange capacity of nanoparticles. The quantity of nano MoS_2 in base oil is limited due to an agglomeration problem and decay in lubrication property

DOI: 10.1201/9781003328742-9

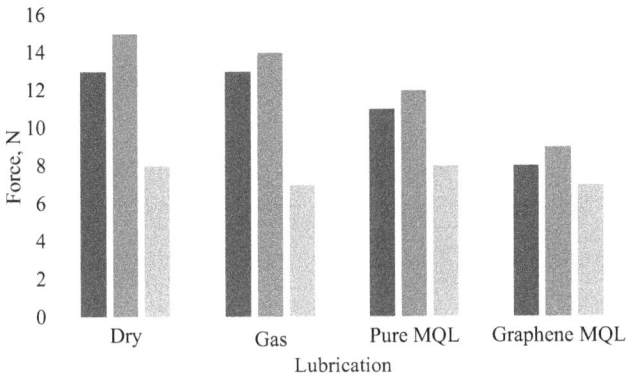

FIGURE 9.1 The average of the positive and negative peak of measured milling force (Li et al., 2018).

(Zhang et al., 2015a). In drilling and tapping operations of ADC12 aluminum alloy 0.5 wt% of graphene concentration exhibited the lowest average torque. The average torque decreased by 11.4%, 16.4%, and 17.7% compared to pure castor oil, rapeseed oil, and corn oil, respectively (Ni et al., 2018). By the introduction of a graphite additive in milling, the vibration intensity reduced due to enhancing the damping effect of oil as the film is formed in the milling zone (Li et al., 2018) (Figure 9.1).

9.1.2 EFFECT ON CUTTING TEMPERATURE

The cutting temperature is reduced significantly with micro- and nano-solid lubricant additives in different base oils. The ability of the solid lubricant to soften at elevated temperatures and exhibit strong adhesion to form a good and consistent film on the metal surface helps to produce a good friction-reducing effect between contact surfaces. The reduced coefficient of friction enables the reduction in heat generation and better penetrability in the tool-work interface (Krishna & Rao, 2008; Padmini et al., 2015). With the addition of solid lubricants in cutting fluid, there is a homogeneous mixing of coolant that results in a metallic surface layer formation that is durable, thick, and strong. This helps to increase the heat transfer coefficient and thus decreases the cutting temperature (Vamsi Krishna et al., 2012). The concentration of nano-particles added to the base fluid influences the quantity of reduction in cutting temperature, as the content of nano-particles increases the thermal conductivity of fluid increases (Srikant et al., 2009). Low tool tip interface temperature is possible with a high amount of solid lubricant nanoparticles in the base fluid as it increases the thermal conductivity of the fluid (Singh et al., 2020). The surface temperature peak and the subsurface temperature peak are significantly reduced due to the solid lubricant additives that enhance the cooling capacity of the oil film formed in the machining zone based on the theory of solid-enhancing heat transfer (Li et al., 2018). Cutting temperatures are compared between dry machining and the coconut oil lubricant added with solid lubricant (boric acid) at different

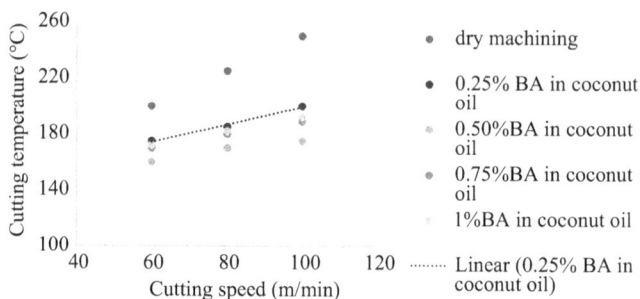

FIGURE 9.2 Variation of cutting temperatures with cutting speed at feed = 0.2 mm/rev, Doc = 1 mm, time = 5 min) (Vamsi Krishna et al., 2012).

proportions *viz.* 0.25%, 0.5%, 0.75%, 1% while machining AISI 1040 steel using a carbide tool, as is illustrated in Figure 9.2 (Vamsi Krishna et al., 2012). Nano and micro-solid lubricants in the base oil in turning AISI 1040 steel and the temperature reduction of 37% and 11%, respectively, are observed, compared to dry machining. Similarly, when nano MoS_2 was compared with dry machining having coconut and sesame oils separately in solid lubricant, tool temperature is reduced by 37% and 31%, respectively. In the case of nano H_3BO_3 and MoS_2, the reduction is 8%, 15% when compared to micro H_3BO_3 and MoS_2 (Padmini et al., 2015). The difference between the surface temperature peak and the subsurface temperature peak in milling of the TC4 alloy under the gas condition, pure MQL, and graphene MQL conditions are 21.47, 15.18, and 7.40°C, which is 6.12%, 33.62%, and 67.64% less than the dry condition (Li et al., 2018).

9.1.3 EFFECT ON SURFACE ROUGHNESS

Surface roughness reduces significantly with solid lubricant additives in a base oil in MQL. This is attributed to the reduction in the coefficient of friction near the tool-work interface due to the formation of lubricating film and thus decreasing the tool wear that helps to improve surface quality (Krishna & Rao, 2008). It is to be noted that the adhesion capability of the solid lubricant plays important role in a surface finish at varying cutting speeds (Sayuti et al., 2013). For example, boric acid gives a better performance due to easy smearing action between the surfaces and nano MoS_2 reduced surface roughness by 46% compared to nano boric acid at a higher temperature due its smearing action at a high temperature (Vamsi Krishna et al., 2010; Padmini et al., 2015). In addition, particle size of the solid lubricant is significant in reducing the surface roughness. A better surface finish is possible with reduced particle size. For example, boric acid solid lubricant with a particle size of 50 µm in turning of EN 8 steel compared with dry, wet, and different particle sizes of graphite and boric acid is shown in Figure 9.3. When the comparison was done with nano and micro-lubricants, the reduction of surface roughness was 13% and 23% for a nano-based boric acid and MoS_2 solid lubricant, respectively (Rao & Krishna, 2008; Padmini et al., 2015). When turning of AL6061-T6 alloy with a MoS_2 solid lubricant, a good surface finish is obtained at 0.5 wt% in oil (Rahmati et al., 2014). In the milling

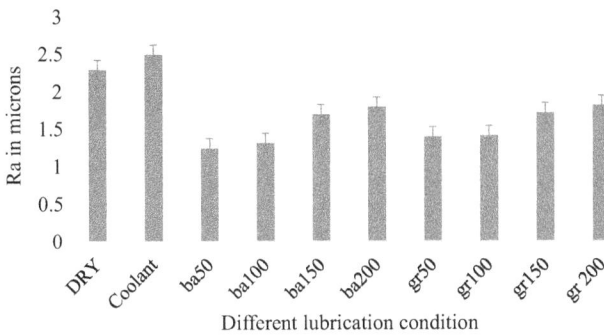

FIGURE 9.3 Average surface roughness (Ra) for different cases (Rao & Krishna, 2008).

of a TC4 alloy, surface roughness is less with a graphene solid lubricant than a dry condition (Li et al., 2018).

The surface finish is also influenced by the workpiece material in a solid lubricant with MQL machining. For example, bearing steel with high hardness (HRC 60) gave a better surface finish compared to medium carbon steel (HRC 22) when machined with graphite solid lubricant. The reason is when grinding is performed on medium carbon steel, the chip wears quickly and combines with graphite and forms a graphite paste that further prevents the escape of the chip from the cutting zone due to an overall increase in wheel load (Shaji & Radhakrishnan, 2002).

9.1.4 EFFECT ON TOOL LIFE AND TOOL WEAR

Solid lubricant suspensions in the base oil with MQL helps in solid lubricant flow at the interface of the tool and workpiece; the direct plastic contact at the tool decreases and reduces the tool wear (Vamsi Krishna et al., 2012). Flank wear decreases with a solid lubricant suspended base oil in MQL compared to other conditions like dry and wet machining due to the formation of lubricant layer on the metal surface (Krishna & Rao, 2008). Low coefficient of friction, sliding action, and low shear resistance within the contact interface help to reduce flank wear because of the self-lubricating film formed by the lattice layer structure of solid lubricants and the reduced frictional effects at the tool and workpiece interaction (Jianhua et al., 2006). Hence, there is a significant reduction of tool wear and increase in tool life with the use of solid lubricant suspensions in a base oil as MQL in machining.

The type of solid lubricant influences the performance of MQL in a machining zone. Boric acid has a better performance in the reduction of tool wear compared to graphite solid lubricant due to its penetration capability (Vamsi Krishna et al., 2010). Adhesion, edge chipping, and built-up edge are significantly reduced under the graphene MQL condition due to enhanced cooling and lubrication of the oil by the film formation in the machining zone (Li et al., 2018). The solid lubricant particle size plays a vital role in the tool wear phenomenon in MQL. For example, in turning AISI 1040 steel with micro and nanosolid lubricants, such as boric acid (H_3BO_3) and molybdenum disulfide (MoS_2), mixed with coconut and sesame oil, an

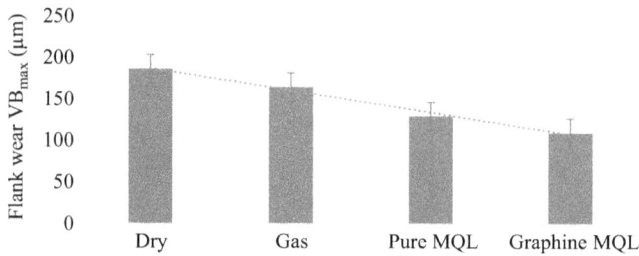

FIGURE 9.4   Maximum flank wear under four lubrication conditions (Li et al., 2018).

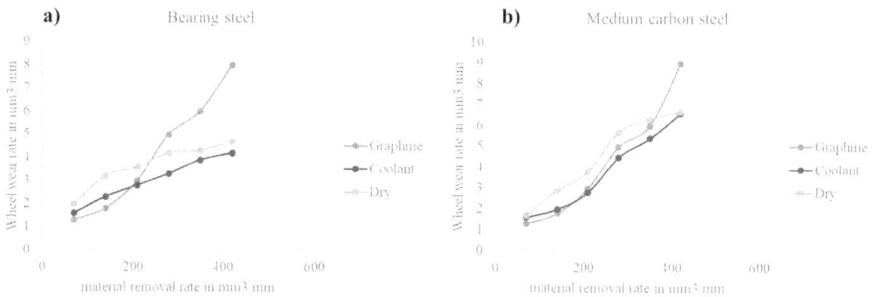

FIGURE 9.5   Volume of wheel worn per unit width with respect to the volume of material removed per unit width under different environments: a) bearing steel, b) medium carbon steel (Shaji & Radhakrishnan, 2002).

average of 10% more reduction in tool wear is observed in a nano-solid lubricant compared to a micro-solid lubricant. The tool flank wear is less due to surface area enhancement with reduced particle size and causes a barrier at the tool work interface, reducing the tool temperature and thus further reducing the flank wear (Padmini et al., 2015). MQL milling of TC4 alloy with four different machining environments, namely dry, gas, pure MQL, and MQL-nanoparticle with graphite reported tool wear is shown in Figure 9.4. The maximum tool flank wear is identified under dry conditions and it is reduced by 6.42%, 25.13%, and 31.02%, respectively, with gas, pure MQL, and graphene MQL compared to dry conditions (Li et al., 2018). In grinding, during the initial stage, the wheel wear is less in graphite-based solid lubrication compared to dry and coolant lubrication. But as machining continues, the wheel wear starts increasing in solid lubrication. This is due to filling up the porous gaps in the grinding wheel with chip and graphite particles, which causes a reduction in sharpness of the grinding wheel, as shown in Figure 9.5 (Shaji & Radhakrishnan, 2002).

9.1.5   Effect on chip morphology

The chip thickness ratio is comparatively less with solid lubricants in MQL compared to dry and wet lubrication due to early smearing action (Vamsi Krishna et al., 2010). This indicates the high shear angle and hence low cutting force in the process. From

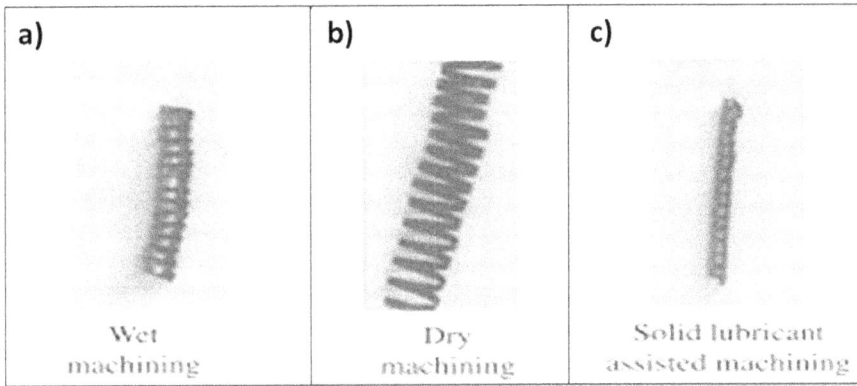

FIGURE 9.6 Comparison of chips generated: a) wet machining, b) dry machining, and c) solid lubricant-assisted machining (Reddy et al., 2010).

Figure 9.6, in drilling operations on AISI 4340 steel with dry, wet, and electrostatic solid lubrication, graphite particle setup with high velocity and low flow rate, the shape of the chips under solid lubricating conditions are short, tabular, and light in color compared to machining under dry and wet lubrications, which indicate solid lubrication causes low heat generation and low friction generation at the tool chip interface (Reddy et al., 2010). The size and mass of solid lubricants influences the chip formation. In turning AISI 1040 steel under dry, wet (flood cooling), and solid lubricant (graphite, boric acid) with particle size 50, 100, 150, and 200 μm and emulsion percentages of 10%, 20%, and 30%, boric acid with a particle size of 50 μm and 20% solid lubricant in oil resulted in a low chip thickness ratio (Vamsi Krishna et al., 2010). When milling AISI 1045 steel with MoS_2 and graphite particles, the chip thickness ratio is better in MoS_2 compared to graphite due to low friction generation because of the presence of sulfur (Reddy & Rao, 2006). The variation of chip thickness ratio at a different cutting speed, feed, radial rake angle, and nose radius at different lubrication conditions such as wet, graphite, and MoS_2, solid lubricant gave good results compared to wet machining, as shown in Figure 9.7.

9.2 HEAT TRANSFER ADDITIVES

There are many types of nanoparticles, such as Al_2O_3, SiO_2, ZrO_2, CuO, TiO_2, CNT, nanodiamonds, and so on, that are proven to be used for improving the thermal conductivity and heat transfer rate of the base fluids in MQL. The parameters of MQL (air pressure and flow rate) and nanofluid (base fluid, the type, size, and concentration of nanoparticles) are influencing the performance in the machining process. Machining of materials such as AISI 1040 steel, AISI 1045, GH4169, Nicrofer C263, Al 6061, and $60Si_2Mn$ steel with MQL use different nanoparticles in order to improve the machining performance. CuO, CNT, Al_2O_3, nanodiamond, nanographite additives, etc. are employed to make nanofluids from base cutting fluids and are successfully used in different machining processes like turning, grinding, milling, and drilling.

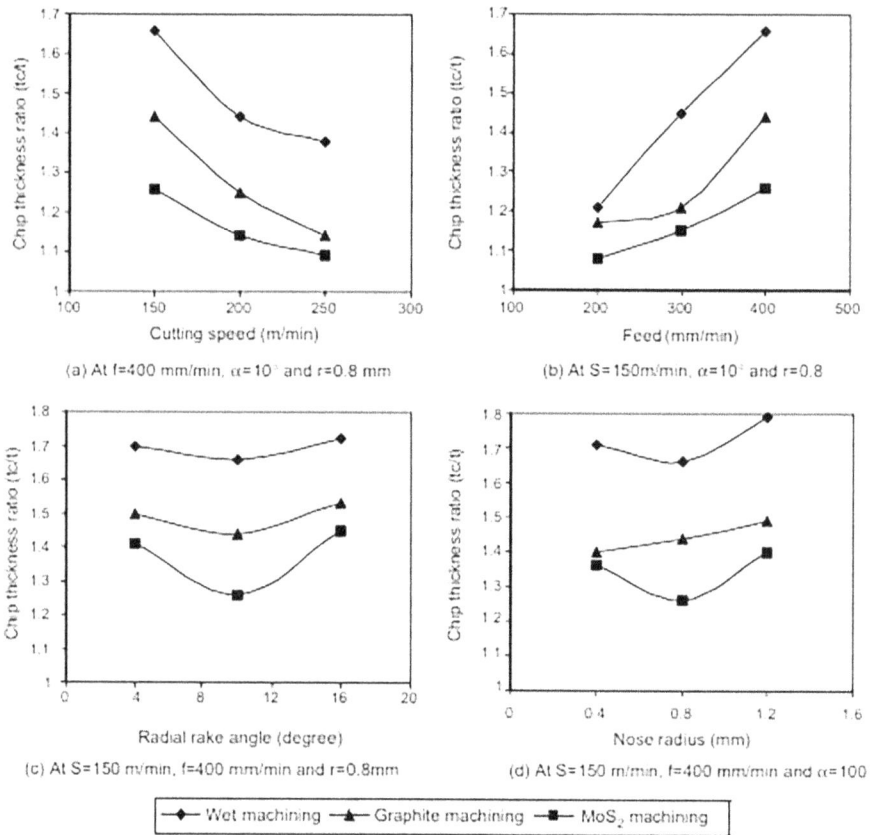

FIGURE 9.7 Variation of chip thickness ratio with a) cutting speed, b) feed rate, c) radial rake angle, d) nose radius (Reddy & Rao, 2006).

9.2.1 EFFECT ON CUTTING FORCE

Fine aerosolization of coolant in MQL by pressurized air can effectively enter the chip-tool interface. This coolant with micro- or nanolubricant additives can form the stable lubrication film between the interacting surfaces and reduces the coefficient of friction. The ball and roller effect of nanoparticles also helps to enter into the cutting zone and friction can be reduced. Hence, inclusion of micro- and nanoparticles in the base oil reduces the cutting force by a change in tool tip interaction, retention of cutting-edge sharpness, and decrease in the formation of a built-up edge. The type of additive and concentration of additives in a base fluid influences the change in cutting force. For example, the lowest cutting force is achieved when 4% of Al_2O_3 is used for turning of Nicrofer C263 compared to a dry, base fluid and 1% of Al_2O_3 (Jawale et al., 2014). A decrease in cutting force is more pronounced with an increase in concentration of nanographite while turning AISI 1040 steel with different cutting tools, as depicted in Figure 9.9. The nanocutting fluid recorded 59%, 29%, and 28.6% lower force compared to dry machining, wet, and mist machining with conventional cutting

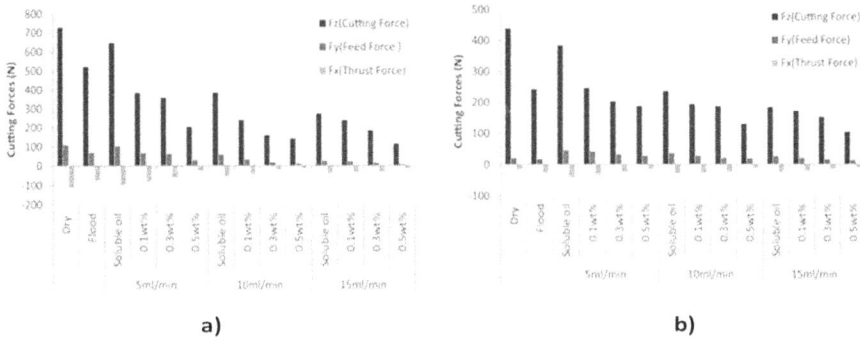

a)

b)

FIGURE 9.8 Variation of cutting forces under varying cutting conditions: a) HSS and b) cemented carbide tool (Amrita et al., 2013).

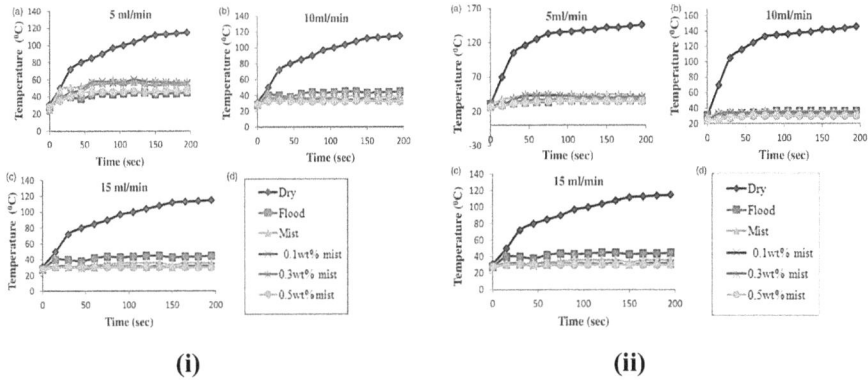

(i)

(ii)

FIGURE 9.9 Variation of chip-tool interface temperature with machining time, i) cemented carbide tool and ii) HSS tool at mist flow rates of: (a) 5 mL/min, (b) 10 mL/min, (c) 15 mL/min, and (d) legend for all cases (Amrita et al., 2013).

fluid, respectively (Sharma et al., 2016). The nanodiamond volumetric concentration of 1% in paraffin oil and that of 2% in vegetable oil are effective for reduction in drilling torque and thrust force (Nam et al., 2011). In some cases, the cutting force seems to be the same for all cooling conditions but after machining time of 15 to 20 minutes, cutting fluid with nanoparticle additives gives a better result in terms of force reduction (Minh et al., 2017) (Figure 9.8).

9.2.2 EFFECT ON CUTTING TEMPERATURE

Micro and nanofluid enhance lubrication in the tool-chip interface by stable film formation while reducing the coefficient of friction. Similarly, these heat transfer additives improve the cooling action in the machining zone due to their enhanced thermal conductivity and heat transfer capacity. Hence, micro and nanofluid with heat transfer additives help in reducing the cutting temperature.

For example, cutting temperature is less with a mist application of nanographite-soluble oil in the turning of AISI 1040 steel due to an increase in thermal conductivity of soluble oil with a nanographite inclusion. The chip-tool interface temperature for a mist application is influenced by wt.% nanoparticles in the base oil and also the flow rate of the fluid in the mist application, as shown in Figure 9.9 (Amrita et al., 2013). In the turning of Nicrofer C263 under the MQL condition, the cutting zone temperature is reduced by 3 to 18% while the addition of nanofluids (4% Al_2O_3) reduced this cutting temperature by 13 to 24% (Jawale et al., 2014).

9.2.3 EFFECT ON SURFACE ROUGHNESS

Different types of nanoparticles have different lubrication mechanisms because of their varying physical properties and thus result in different lubrication effects. Nanofluids exhibit enhanced thermal properties, such as higher thermal conductivity and heat transfer coefficients, and also greatly improve the wetting and lubricating properties on the rake face of the tool and can be helpful in improving the surface finish. Surface roughness improves with MQL compared to dry cutting and this can be further enhanced using nanofluids in MQL. For example, Al_2O_3 nanoparticles from 1–4% in soluble oil improved surface integrity while using MQL (Jawale et al., 2014). The performance of these micro/nanofluids in MQL is influenced by particle size and percentage of particles, as discussed in the previous sections. In one of the case studies, enhancement in the surface quality in micro-scale grinding is observed with a 30 nm size and 2% volumetric concentration for both nanodiamond (64.3%) and nano Al_2O_3 (65.5%) particles compared to dry air, as shown in Figure 9.10. This is because smaller particles produce smoother surfaces than the larger ones, as smaller nanoparticles could be absorbed more in the grinding area, and their ball-bearing effect might be bigger (Lee et al., 2012).

With Al_2O_3 nanofluid, the surface roughness is reduced by 47.8%, 29.1%, and 25.5%, compared to dry machining, conventional mist, and wet machining, respectively (Sharma et al., 2016). Nanofluid MQL can be an effective solution to easily remove chips and burrs during the micro-drilling process, which also improves the surface finish of drilled holes (Nam et al., 2011). In hard milling under MQL conditions with 0.5% Al_2O_3 nanofluid, surface roughness is reduced by 66%. Sometimes the performance of nanoparticles may not be really clear, but after a certain period of time, due to the oil film with nanoparticles formed in the contact zone, it creates "roller effect" that plays an important role. Rolling friction instead of sliding occurs between the flank face and machined surface using nanofluids in MQL. However, in MQL cutting with base fluids, rapid tool wear occurs and thus surface roughness increases, as shown in Figure 9.11 (Minh et al., 2017).

9.2.4 EFFECT ON TOOL WEAR

Adhesion of the material on the tool surface followed by abrasive wear mechanism is the main reason for the tool flank wear. Notching occurs at the tool nose, caused by adhesion due to the generation of high temperature and pressure between the cutting tool and workpiece that leads to plastic deformation in contact area.

(a) Nanodiamond particles

(b) Nano-Al_2O_3 particles

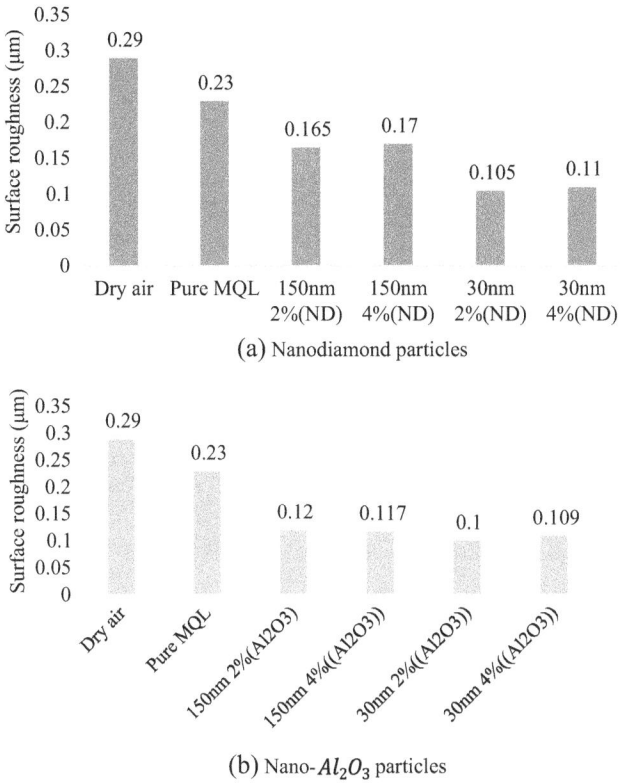

FIGURE 9.10 Measured surface roughness (Ra) of ground workpieces in the cases using a) nanodiamond particles and b) nano-Al_2O_3 particles (Lee et al., 2012).

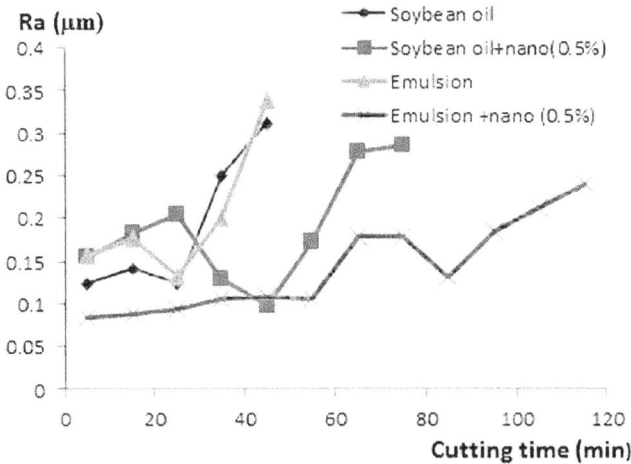

FIGURE 9.11 Measured surface roughness (Ra) in milling of $60Si_2$Mn steel under MQL conditions with and without nanofluid (Minh et al., 2017).

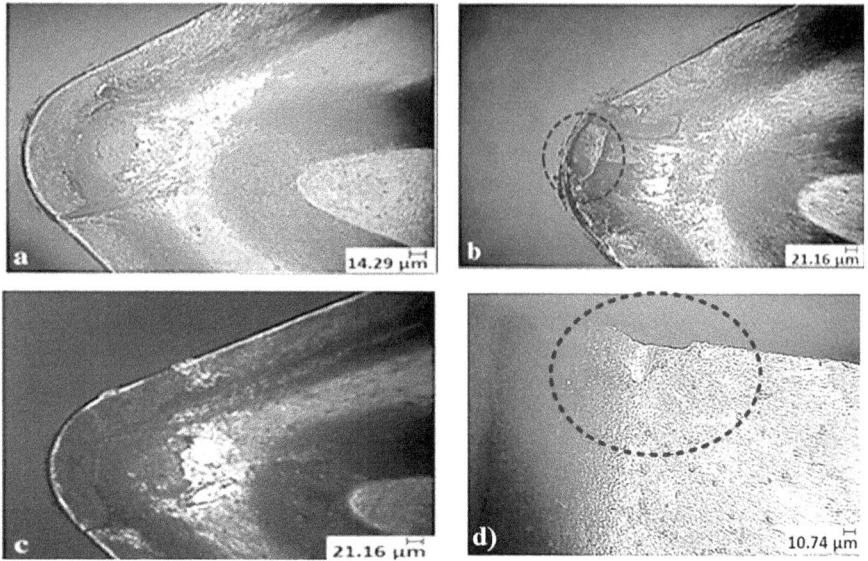

FIGURE 9.12   Crater wear at cutting speed of 60 m/min and feed rate of 0.2 mm/rev. (a) 4 wt.% Al_2O_3 nano-fluid, (b) without nanoadditives, (c) 4 wt.% MWCNTs nano-fluid, and (d) notch wear when using 4 wt.% Al_2O_3 nanofluid at cutting speed of 60 m/min and feed rate of 0.3 mm/rev (Hegab et al., 2018).

Nanofluids offer an effective heat dissipation performance that retains the cutting tool original hardness. Using nanofluids, cooling and lubrication properties are improved, which enhance the interface bonding between the tool and workpiece surfaces and, as a result, the adhesive wear severity can be partially eliminated and accordingly, the flank wear will be reduced (Hegab et al., 2018). Nanoadditives work as spacers between the contact surfaces and reduce the friction between the tool-chip interface. Hence, nanoadditives show a significant improvement in reducing the flank wear. For example, the crater wear with Al_2O_3 nanoadditives and without nano-additives is shown in Figure 9.12(a) and (b). The tool wear with Al_2O_3 and MWCNT with same concentration in the base fluid is observed in Figure 9.12(a) and (c). The notch wear with a Al_2O_3 nanofluid is shown in Figure 9.12(d).

Flank wear is reduced with an increase in the percentage of nanoparticles in soluble oil due to enhanced thermal conductivity of soluble oil with the inclusion of nanographite or due to the lubrication effect of nanographite or due to the combination of both, as seen in Figures 9.13, 9.14, and 9.15, respectively (Amrita et al., 2013). Flank wear is reduced with an increase in the mist flow rate by increasing the air pressure. The aerosol droplet size decreases with air pressure and effectively enters the tool-chip interface, causing efficient cooling. It is also identified that the mist flow rate also plays an important role in reducing the cutting zone temperature and enhancing the tool life. For example, while machining with the HSS tool in a mist application, oil breaks into fine aerosols that can easily enter the chip-tool

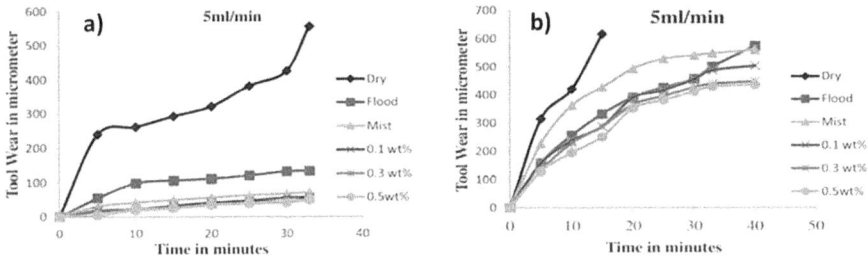

FIGURE 9.13 Variation of flank wear with machining time at 5 mL/min: a) cemented carbide tool and b) HSS (Amrita et al., 2013).

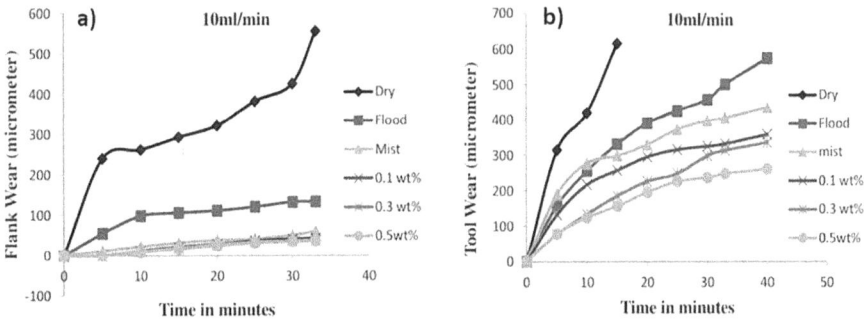

FIGURE 9.14 Variation of flank wear with machining time at 10 mL/min: a) cemented carbide tool and b) HSS (Amrita et al., 2013).

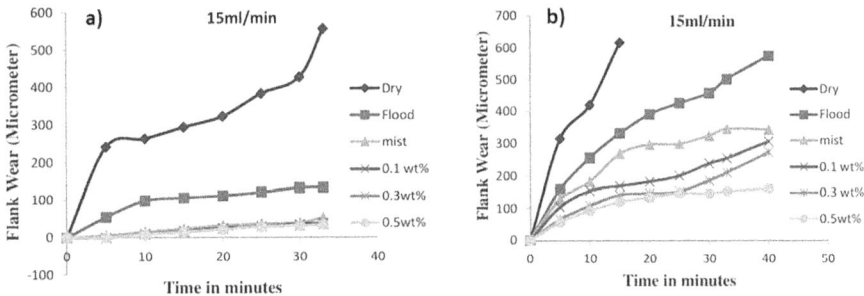

FIGURE 9.15 Variation of flank wear with machining time at 15 mL/min: a) cemented carbide tool and b) HSS (Amrita et al., 2013).

interface region and remove the generated heat effectively and thus reduce the tool wear efficiently compared to dry and flood lubrication.

The reduction in tool wear is because of diffusion and penetration of nanofluid mist into machining zone and reduced frictional force. The lower frictional force reduces the cutting temperature due to a ball bearing effect of nanoparticles present in the cutting fluid. The tool wear is less with Al_2O_3 nanofluids by 63.9%, 44.9%,

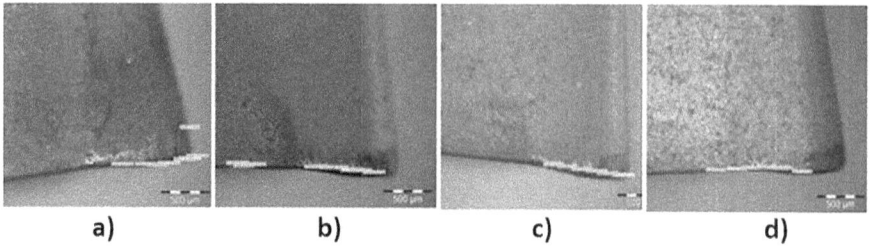

| a) | b) | c) | d) |

FIGURE 9.16 Microscopic photographs of tool flank wear under different machining conditions: a) dry, b) conventional cutting fluid mist, c) conventional wet, and d) 1% Al_2O_3 nanofluid mist (Sharma et al., 2016).

and 5.27%, compared to dry, conventional mist, and wet machining, respectively, as shown in Figure 9.16 (Sharma et al., 2016).

9.2.5 EFFECT ON TOOL LIFE

Formation of oil film with nanoparticles in the contact zone plays an important role in creating a roller effect. Rolling friction occurs between the flank face and machined surface, rake face and chip surface, and so forth. Hence, cutting forces, cutting temperature, tool wear, and surface roughness are reduced with the application of nanofluids in MQL and thus helps to improve the tool life. Also due to the ball bearing effect, burr elimination and easy removal of chips occurs. For example, in the drilling of Al 6061 with the pure and nanofluid MQLs, the microdrills are used for making 150 holes without failure due to their enhanced cooling and lubrication effects and reduced chip adhesion to the microdrill, while the microdrill failed at the 87th hole in compressed air lubrication (Nam et al., 2011). In hard milling of $60Si_2$Mn steel with soybean oil mixed with nano Al_2O_3, the tool life increased compared to pure soybean oil. In the case of 5% emulsion coolant with nano Al_2O_3, the tool life increased compared to pure 5% emulsion coolant, as shown in Figure 9.17 (Minh et al., 2017). However, the use of emulsion in MQL is

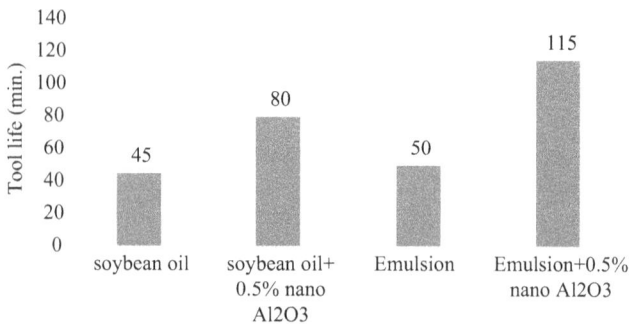

FIGURE 9.17 Tool life under MQL conditions with or without nanofluids (Minh et al., 2017).

more effective than that of pure soybean oil. Hence, tool life in the case of emulsion is higher than that of soybean oil.

9.2.6 EFFECT ON CHIP MORPHOLOGY

Chip morphology provides important clues about the cutting mechanics and is closely related to surface integrity of the finished product (surface roughness, surface topography, etc.) and machining efficiency. In the turning of Inconel 718, using MWCNT's nanofluid showed tiny segmented chips because of the impingement of MWCNTs on the generated chips, which increased the chip helix angle and forced the long chips to break. The chip morphology under different conditions is shown in Figure 9.18. In addition, the BUE effects on the generated chips at a cutting speed of 60 m/min and feed rate of 0.3 mm/rev can be seen without the nanoadditives condition. BUE formation does not take place in MWCNTs or Al_2O_3 nanofluids due to their promising cooling and lubrication properties.

Dry machining produces blue chips due to the high temperatures in the shear and cutting zones. Using conventional cutting, fluid mist reduces the temperature in the machining zone creating slightly golden chips. Under flood lubrication, the chip color changed to shiny silver which shows the temperature in the treatment zone has decreased further. The application of nano cutting fluid mist during processing of silvery chips may be due to the enhanced cooling and lubricating effect of nanoparticles present in the nanofluid, as shown in Table 9.1 (Sharma et al., 2016).

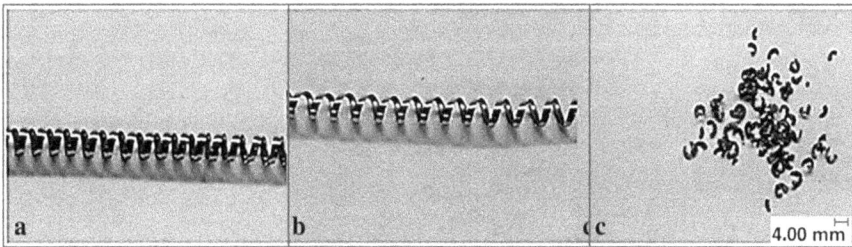

FIGURE 9.18 Chips generated during machining of Inconel 718 at a cutting speed of 60 m/min and feed rate of 0.3 mm/rev: (a) without nanoadditives, (b) Al_2O_3 nanofluid, and (c) MWCNT's nanofluid (Hegab et al., 2018).

9.3 HYBRID ADDITIVES

Different cutting fluids are used in MQL by adding the hybrid additives to the base fluid to enhance the machining performance. These hybrid nanofluids are used in the machining of steels, titanium, and nickel alloys such as AISI 1040 steel, Inconel X-750 alloy, and GH4169 Ni-based alloy under MQL with hybrid nanoparticles such as hBN/Gr, hBN/MoS_2, Gr/MoS_2, CNT/MoS_2, SiO_2-Al_2O_3-ZrO_2 with different compositions. The variation of cutting force, cutting temperature, surface roughness, tool wear, and chip morphology using such additives in MQL are discussed below.

TABLE 9.1

Summarizes the shape and color of chips produced during turning in different machining environments (Sharma et al., 2016)

Machining Environment	Chip Photo	Chip Shape	Chip Color
Dry		Curl and segmented	Blue
Wet		Curl and segmented	Silver
MQL with conventional cutting fluid		Curl and segmented	Light Golden
MQL with nanofluid		Curl and segmented	Silver

9.3.1 Effect on cutting force

Many parameters, such as tool geometry, tool and workpiece properties, cutting parameters, coefficient of friction, cooling and lubrication conditions, affect the cutting force. The cutting force is reduced by using cutting fluids that easily flow into the tool-work interface and reduce friction. Hybrid nanofluids with MQL with low viscosity easily penetrate into the tool-work interface, act as a lubrication layer,

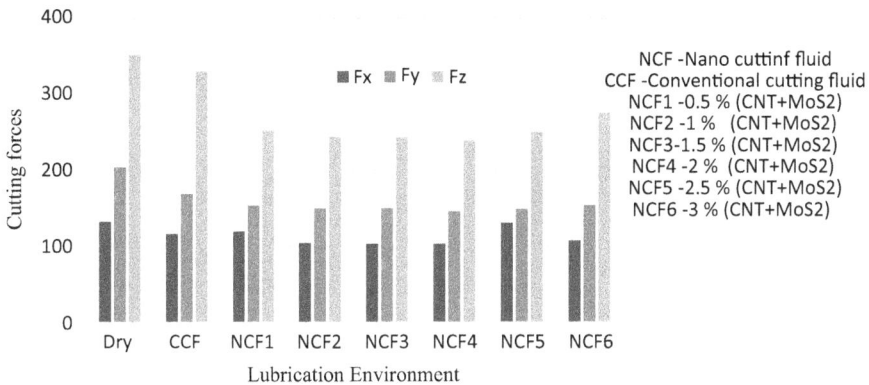

FIGURE 9.19 Variation of cutting force at different lubrication environments (Gugulothu & Pasam, 2022).

reduce the friction, and thus decrease the cutting forces (Şirin & Kıvak, 2021). The mixing of two nanoparticles in a base fluid shows more effect on machining performance compared to a single base nanofluid because of synergistic behavior and better lubrication. The concentration of additives in a nanofluid also shows variation in cutting forces. However, as the concentration increases, after a certain level the force also increases. At higher concentrations, nanoparticles settle at the machining zone and form a dense layer and increase the cutting force. Viscosity is the main reason for formation of the dense layer on a workpiece and this layer obstructs the tool movement, which leads to an increase in cutting force (Gugulothu & Pasam, 2022). Figure 9.19 shows the cutting force variation with nanofluid, conventional cutting fluid, and different concentrations of nanofluid. Cutting forces are also influenced by size and concentration of nanoparticles in the base fluid (Zhang et al., 2016).

9.3.2 Effect on Cutting Temperatures

Heat is generated during machining due to plastic deformation in shear zone, friction between the tool and workpiece, as well as the chip and the tool. Cutting temperature depends on cutting parameters, type of material, cooling conditions, and so on. Different types of nanofluids are prepared with the addition of two or more nanoparticles and are used for enhancing the heat transfer properties. Thermal conductivity is the most important thing in heat dissipation from the tool-work interface. The use of high thermal conductivity particles improves the heat dissipation at the cutting zone. Thermal conductivity and viscosity also show an impact on the cutting temperatures. The use of low viscosity and high thermal conductivity nanofluid promotes low friction between contact surfaces, which helps to reduce heat generation to some extent, dissipate more heat from the machining zone quickly, and thus helps to reduce the cutting temperature (Şirin & Kıvak, 2021). High concentrations of nanoparticles possess superior thermophysical properties to absorb and rapidly carry the heat away from the cutting

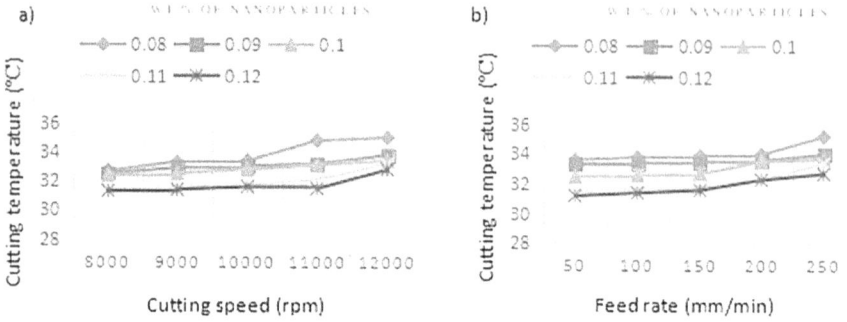

FIGURE 9.20   Cutting temperatures with respect to speed and feed rate (Safiei et al., 2021).

zone. These nanoparticles have good thermal conductivity and hence can dissipate the heat to the surroundings (Safiei et al., 2021). Nanoparticles of CNT between the MoS_2 particles form a sandwich shape and generate a sliding motion that reduces the friction that further decreases the cutting temperatures (Gugulothu & Pasam, 2022). Cutting temperature is reduced at constant and even at higher cutting parameters with the use of hybrid nanofluids (Geetha et al., 2021). For example, Figure 9.20 shows the cutting temperature with respect to cutting speed and feed rate with use of SiO_2-Al_2O_3-ZrO_2 tri-hybrid nanofluid. In any machining operation, the use of hybrid nanoparticles under MQL conditions reduces the cutting temperatures drastically.

9.3.3   EFFECT ON SURFACE ROUGHNESS

The output responses, such as cutting temperatures and tool wear, affect the surface roughness directly. By using hybrid fluids, these parameters are reduced and thus the surface quality of the component is improved. Hence, application of MQL with hybrid nanofluids in any machining process reduces the surface roughness. Nanoparticles of spherical shape play a ball rolling effect on the friction surface and reduce the coefficient of friction that leads to a decrease in surface roughness of the machined component (Şirin & Kıvak, 2021). Surface roughness is more influenced by the feed and also depends on heat dissipation properties of nanoparticles present in nanofluids. The thermal conductivity of nano-fluid depends on thermal conductivity of nanoparticles and Brownian motion of particles (Gugulothu & Pasam, 2022). Figure 9.21 shows the surface roughness with the use of a hybrid nanofluid prepared with different concentrations (Zhang et al., 2016). Even though nanofluids help to reduce surface roughness, sometimes an increase in the nanoparticle concentration beyond a limit leads to dominance of repulsive force between the nanoparticles, which tend to aggregate the nanoparticle and provide poor lubrication. Hence, proper choice of nanoparticle concentration in the base fluid is necessary for yielding a better surface finish (Gugulothu & Pasam, 2022). The coefficient of friction for four hybrid nanofluids (prepared by varying the concentration of MoS_2 and CNT) are lower than those of pure MoS_2 and pure CNT nanofluids, reflecting their better

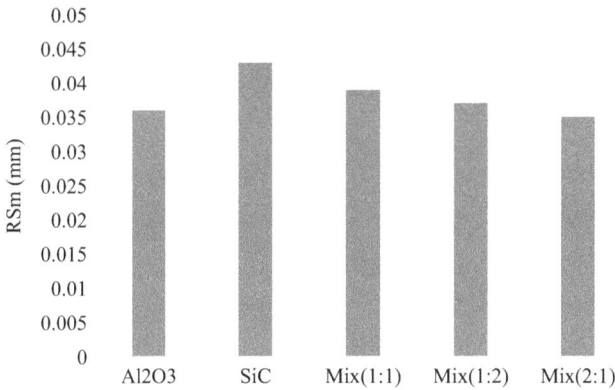

FIGURE 9.21   Surface roughness of different Al$_2$O$_3$/SiC mixing ratios (Zhang et al., 2016).

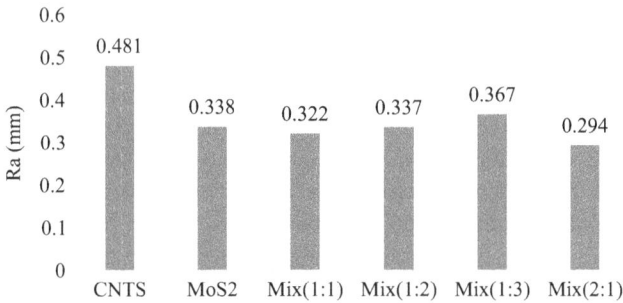

FIGURE 9.22   Workpiece surface roughness (Ra) of MQL grinding using different nano-fluids (Zhang et al., 2015b).

lubrication effect. Using pure CNT nanofluids produces a larger "furrow" on surface, resulting in poor workpiece surface quality. As shown in Figure 9.22, Ra for mix (2:1) is the lowest of six nanofluids and is 13% and 39% lower than MoS$_2$ and CNTs, respectively, while other mixture has slightly less value compared to the mix (2:1) (Zhang et al., 2015b). Thus, proper nanoparticle concentration is also important in hybrid nanofluids to reduce the surface roughness of the machined component.

9.3.4   EFFECT ON TOOL WEAR AND TOOL LIFE

The addition of nanoparticles with high thermal conductivity helps to increase the conductivity of the base fluid. By using these hybrid nanofluids in MQL, it gives lower cutting temperatures, which leads to a reduction in tool wear and increases the tool life (Geetha et al., 2021). For example, flank wear is reduced by 22% while turning AISI 4030 with the use of (Cu-Gr) nanofluid under MQL compared to dry machining conditions, as shown in Figure 9.23. There is improvement in tribological and thermal properties with 1.5 wt% of Al-SiC/soluble oil hybrid

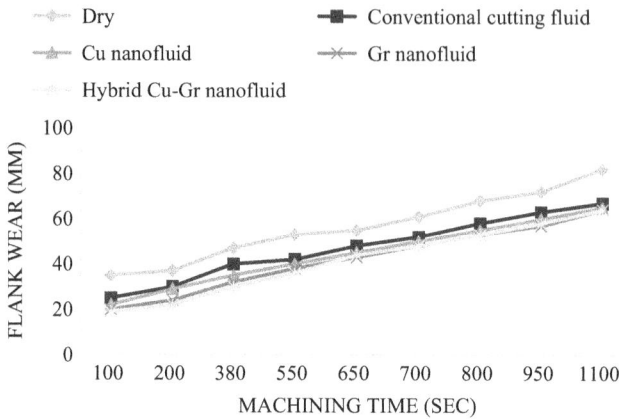

FIGURE 9.23 Variation of flank wear with respect to time (Geetha et al., 2021).

nanofluids in machining EN-24 steel and decreases the tool wear (Thakur et al., 2022). Higher concentration of nanoparticles in hybrid nanofluid lead to a decrease in tool wear due to lubrication and cooling. Compared to other conditions, hybrid nanofluids with MQL give better machining efficiency and performance. The use of hybrid additives with MQL can improve the overall tool life by reducing wear and tear on the tool. This can be particularly beneficial when machining titanium, which is known for its high strength and toughness.

9.3.5 EFFECT ON CHIP MORPHOLOGY

Chip morphology, such as type and color of chips, mostly depends on the machining conditions, workpiece and tool material, cutting force, cutting temperature, lubrication, and cooling conditions. By application of MQL with hybrid nanofluid, the chips are continuous and the surface finish on the component is also improved. Gold-colored chips are formed while turning AISI 4030 under dry conditions due to high cutting temperatures, whereas hybrid (Cu-Gr) nanofluid with MQL produced silver-colored chips due to a reduction in cutting temperatures (Geetha et al., 2021). Sometimes discontinuous chips are also formed due to this cooling effect of nanofluids. The high temperatures soften the chip and leads to the formation of squeezed string deleterious chips. By applying hybrid nanofluid, the quality of the chip is improved. The nanoparticles added to the base oil increase the thermal conductivity of fluid and enhance the heat dissipation from the tool workpiece interface and form the segmented small ribbon-shaped chips under optimum cutting conditions, as shown in Figure 9.24 (Singh et al., 2021). This type of chip depends on a concentration of nanoparticles that provides low friction and high thermal conductivity to fluid. While turning the grade 5 titanium alloy with graphite-talc hybrid nanofluid in MQL, the segmented type of chip is formed. Hybrid nanofluids are mostly preferred to get good machining performance, considering chip formation.

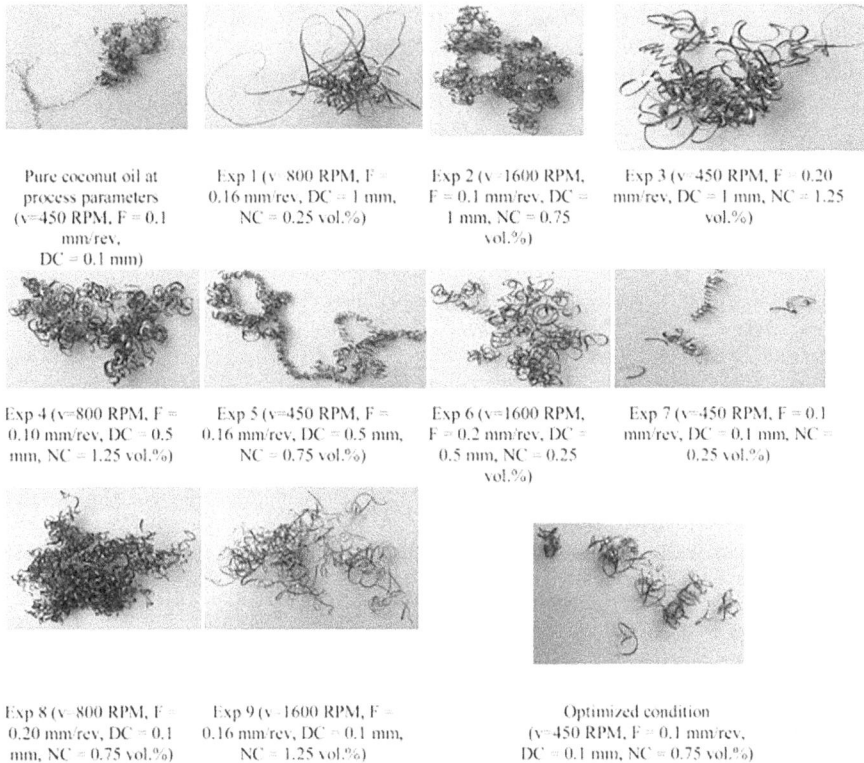

| Pure coconut oil at process parameters (v=450 RPM, F = 0.1 mm/rev, DC = 0.1 mm) | Exp 1 (v = 800 RPM, F = 0.16 mm/rev, DC = 1 mm, NC = 0.25 vol.%) | Exp 2 (v = 1600 RPM, F = 0.1 mm/rev, DC = 1 mm, NC = 0.75 vol.%) | Exp 3 (v = 450 RPM, F = 0.20 mm/rev, DC = 1 mm, NC = 1.25 vol.%) |

| Exp 4 (v=800 RPM, F = 0.10 mm/rev, DC = 0.5 mm, NC = 1.25 vol.%) | Exp 5 (v=450 RPM, F = 0.16 mm/rev, DC = 0.5 mm, NC = 0.75 vol.%) | Exp 6 (v=1600 RPM, F = 0.2 mm/rev, DC = 0.5 mm, NC = 0.25 vol.%) | Exp 7 (v = 450 RPM, F = 0.1 mm/rev, DC = 0.1 mm, NC = 0.25 vol.%) |

| Exp 8 (v = 800 RPM, F = 0.20 mm/rev, DC = 0.1 mm, NC = 0.75 vol.%) | Exp 9 (v = 1600 RPM, F = 0.16 mm/rev, DC = 0.1 mm, NC = 1.25 vol.%) | Optimized condition (v=450 RPM, F = 0.1 mm/rev, DC = 0.1 mm, NC = 0.75 vol.%) |

FIGURE 9.24 Chip types during different experimental runs (Singh et al., 2021).

REFERENCES

Amrita, M., Srikant, R., Sitaramaraju, A., Prasad, M., & Krishna, V. (2013). Experimental investigations on influence of mist cooling using nanofluids on machining parameters in turning AISI 1040 steel. *Proceedings of the Institution of Mechanical Engineers, Part J: Journal of Engineering Tribology*. 227. 1334–1346. 10.1177/1350650113491934.

Geetha, C. H. T. S., et al. (2021). Analysis of hybrid nanofluids in machining AISI 4340 using minimum quantity lubrication. *Materials Today: Proceedings*, 43, 579–586.

Gugulothu, S., & Pasam, V. K. (2022). Experimental investigation to study the performance of CNT/MoS2 hybrid nanofluid in turning of AISI 1040 stee. *Australian Journal of Mechanical Engineering*, 20(3), 814–824.

Hegab, H., Umer, U., Soliman, M., & Kishawy, H. A. (2018). Effects of nano-cutting fluids on tool performance and chip morphology during machining Inconel 718. *The International Journal of Advanced Manufacturing Technology*, 96(9), 3449–3458.

Jawale, K., & Bose, P. S. C., & Rao, C. (2014). Role of MQL and nano fluids on the machining of Nicrofer C263.

Jianhua, D., Tongkun, C., Xuefeng, Y., & Jianhua, L. (2006). Self-lubrication of sintered ceramic toos with CaF2 additions in dry cutting. *Int. J. Mach. Tools Manuf.*, 46, 957–963.

Krishna, P. V., & Rao, D. N. (2008). Performance evaluation of solid lubricants in terms of machining parameters in turning. *International Journal of Machine Tools and Manufacture*, 48(10), 1131–1137.

Lee, P. H., Nam, J. S., Li, C., & Lee, S. W. (2012). An experimental study on micro-grinding process with nanofluid minimum quantity lubrication (MQL). *International Journal of Precision Engineering and Manufacturing*, 13(3), 331–338.

Li, M., et al. (2018). MQL milling of TC4 alloy by dispersing graphene into vegetable oil-based cutting fluid. *The International Journal of Advanced Manufacturing Technology*, 99(5), 1735–1753.

Minh, D. T., The, L. T., & Bao, N. T. (2017). Performance of Al2O3 nanofluids in minimum quantity lubrication in hard milling of 60Si2Mn steel using cemented carbide tools. *Advances in Mechanical Engineering*, 9(7), 1687814017710618.

Nam, J. S., Lee, P. H., & Lee, S. W. (2011). Experimental characterization of micro-drilling process using nanofluid minimum quantity lubrication. *International Journal of Machine Tools and Manufacture*, 51(7-8), 649–652.

Ni, J., et al. (2018). Reinforced lubrication of vegetable oils with graphene additive in tapping ADC12 aluminum alloy. *The International Journal of Advanced Manufacturing Technology*, 94(1), 1031–1040.

Padmini, R., Krishna, P. V., & Rao, G. K. M. (2015). Performance assessment of micro and nano solid lubricant suspensions in vegetable oils during machining. *Proceedings of the Institution of Mechanical Engineers, Part B: Journal of Engineering Manufacture*, 229(12), 2196–2204.

Rahmati, B., Sarhan, A. A. D., & Sayuti, M. (2014). Investigating the optimum molybdenum disulfide (MoS2) nanolubrication parameters in CNC milling of AL6061-T6 alloy. *The International Journal of Advanced Manufacturing Technology*, 70(5), 1143–1155.

Rao, D. N., & Krishna, P. V. (2008). The influence of solid lubricant particle size on machining parameters in turning. *International Journal of Machine Tools and Manufacture* 48(1), 107–111.

Reddy, N. S. K., & Rao, P. V. (2006). Experimental investigation to study the effect of solid lubricants on cutting forces and surface quality in end milling. *International Journal of Machine Tools and Manufacture*, 46(2), 189–198.

Reddy, N. S. K., Nouari, M., & Yang, M. (2010). Development of electrostatic solid lubrication system for improvement in machining process performance. *International Journal of Machine Tools and Manufacture*, 50(9), 789–797.

Safiei, W., et al. (2021). Effects of SiO2-Al2O3-ZrO2 Tri-hybrid Nanofluids on Surface Roughness and Cutting Temperature in End Milling Process of Aluminum Alloy 6061-T6 Using Uncoated and Coated Cutting Inserts with Minimal Quantity Lubricant Method. *Arabian Journal for Science and Engineering*, 46(8), 7699–7718.

Sayuti, M., Sarhan, A. A. D., & Hamdi, M. (2013). An investigation of optimum SiO2 nanolubrication parameters in end milling of aerospace Al6061-T6 alloy. *The International Journal of Advanced Manufacturing Technology*, 67(1), 833–849.

Shaji, S., & Radhakrishnan, V. (2002). An investigation on surface grinding using graphite as lubricant. *International Journal of Machine tools and manufacture*, 42(6), 733–740.

Sharma, A. K., Singh, R. K., Dixit, A. R., & Tiwari, A. K. (2016). Characterization and experimental investigation of Al2O3 nanoparticle based cutting fluid in turning of AISI 1040 steel under minimum quantity lubrication (MQL). *Materials Today: Proceedings*, 3(6), 1899–1906.

Singh, B. K., et al. (2020). Evaluation of mechanical and frictional properties of CuO added MgO/ZTA ceramics. *Materials Research Express*, 6(12), 125208.

Singh, V., Sharma, A. K., Sahu, R. K., & Katiyar, J. K. (2021). Novel application of graphite-talc hybrid nanoparticle enriched cutting fluid in turning operation. *Journal of Manufacturing Processes*, 62, 378–387.

Şirin, Ş., & Kıvak, T. (2021). Effects of hybrid nanofluids on machining performance in MQL-milling of Inconel X-750 superalloy. *Journal of Manufacturing Processes*, 70, 163–176.

Srikant, R. R., et al. (2009). Applicability of cutting fluids with nanoparticle inclusion as coolants in machining. *Proceedings of the Institution of Mechanical Engineers, Part J: Journal of Engineering Tribology*, 223(2), 221–225.

Thakur, A., Manna, A., & Samir, S. (2022). Performance evaluation of Al-SiC nanofluids based MQL sustainable cooling techniques during turning of EN-24 steel. *Silicon*, 14(3), 869–882.

Vamsi Krishna, P., Srikant, R. R., & Rao, D. N. (2010). Experimental investigation to study the performance of solid lubricants in turning of AISI1040 steel. *Proceedings of the Institution of Mechanical Engineers, Part J: Journal of Engineering Tribology*, 224(12), 1273–1281.

Vamsi Krishna, P., et al. (2012). Basic properties and performance of vegetable oil-based boric acid nanofluids in machining. *Emerging trends in science, engineering and technology*. Springer, India. 197–206.

Zailani, Z. A., et al. (2011). The influence of solid lubricant in machining parameter of milling operation. *International Journal of Engineering Science and Technology*, 3(6).

Zhang, X., et al. (2016). Performances of Al2O3/SiC hybrid nanofluids in minimum-quantity lubrication grinding. *The International Journal of Advanced Manufacturing Technology*, 86(9), 3427–3441.

Zhang, Y., et al. (2015a). Experimental evaluation of MoS2 nanoparticles in jet MQL grinding with different types of vegetable oil as base oil. *Journal of Cleaner Production*, 87, 930–940.

Zhang, Y., Li, C., Jia, D., Zhang, D., & Zhang, X. (2015b). Experimental evaluation of the lubrication performance of MoS2/CNT nanofluid for minimal quantity lubrication in Ni-based alloy grinding. *International Journal of Machine Tools and Manufacture*, 99, 19–33.

10 Variants of MQL

Minimum Quantity Cooling Lubrication

10.1 INTRODUCTION

Minimum quantity cooling lubrication can also be termed *MQCL*. Inclusion of cold air to the MQL constitutes a minimum quantity cooling lubrication method, which significantly improves the performance of the MQL (Maruda et al., 2014). Several techniques, including cryogenic cooling, vortex tubes, and refrigeration systems, are used to create cold air, which is added to the aerosol and directed towards the machining interface. MQL aerosol, which is a combination of lubricating oil and compressed air, lowers friction by resulting in the generation of a thin film of lubrication in between the tool face and chip during machining. As a result, the tool life is eventually increased. But unlike conventional cutting fluids, utilizing the MQL aerosol does not lower the cutting temperature (Su et al., 2007). Therefore, in order to minimize the temperature at the machining interface, cold air must also be given along with the MQL aerosol. The cold air temperature can be at a sub-zero temperature. Different cooling systems can be used based upon the applications and the desired temperature of the cold air. A vortex tube can create cold air up to −4 °C, a refrigeration system can produce up to −45 °C, and cryogenic cooling using liquid nitrogen (LN_2) can create as low a temperature as −196 °C. From these three ways of generating cold air, vortex tubes are claimed to be the simplest and cost-effective technique. MQCL has a different lubrication mechanism to produce cold air with MQL aerosol, which involves single jet and multi-jet MQCL and the use of various types of MQCL depends on the machining process and the performance (Mark Benjamin et al., 2018). The emulsion mist formed by MQCL is key to improving the cooling lubrication in the machining zone, which enhances the efficiency of the machining process. The tribo-film formed on the tool-chip interface results in lower friction and reduced tool wear. The emulsion mist results in a significant amount of heat being dissipated from the machining zone (Maruda et al., 2017). Here, the droplet diameter depends on the distance of the nozzle and is controlled by the volumetric air flow. The results are better in mist formation, in which all the droplets fall on the machining interface and get evaporated within few seconds from the machining interface (Maruda et al., 2016).

MQCL performance can be improved by adding additives to the MQCL base fluid, which is proved by many of the works. Nanoparticles such as graphite, MoS_2, Al_2O_3 MWCNTs, TiO_2, etc. are generally used as the additives to increase the lubrication effect and increase heat dissipation at the machining interface (Okokpujie et al,. 2022; Tuan et al,. 2022). Figure 10.1 shows the TEM image of the nanoparticles, where we can see the size of the nanoparticle is responsible for easier

DOI: 10.1201/9781003328742-10

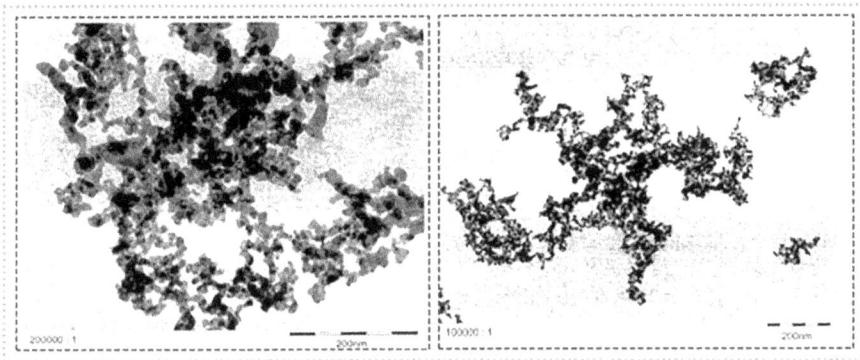

FIGURE 10.1   TEM image of Al$_2$O$_3$ nanoparticles (Pal et al., 2021).

penetration into the machining interface (Pal et al,. 2019). Nanoparticles, when added to the lubricating oil, improve the performance of the lubrication while machining by decreasing friction between the chip and the tool. Nanoparticles settle between the chip and the tool, which results in a rolling action of the chip and helps the chips to easily move from the machining interface.

Nanoparticles, which have high thermal conductivity, help to improve heat dissipation from the machining interface by increasing the heat dissipation capacity of the lubricating oil (Subhedar et al,. 2021). Moreover, application of nanoparticles also promotes sustainability while also improving the performance of the lubricating oil in MQCL (Achparaki et al,. 2012).

10.2   MQCL SETUP AND LUBRICATION MECHANISM

Typically, a minimum quantity lubrication system consists of two primary systems: one for MQL aerosol, which delivers compressed air and lubricating oil, and the other for cold air supplement. To be used efficiently when machining, these two systems are positioned at an angle and at a specific distance from the machining interface. The lubricating oil is assorted with compressed air in MQL system, and the aerosol mixture is delivered from the nozzle. The discharge rate and compressed air pressure both play important roles in the lubricating mechanism. In a cold air supplement system, air is cooled and supplied across the machining zone via a vortex tube or other cooling system. The lubricant oil from the reservoir is discharged with a specified discharge rate with the help of a solenoid pump mixture then flows through the nozzle onto the machining interface in the form of mist. The other part is where the air is compressed to a certain pressure and is passed through the vortex tubes or any certain air-cooling mechanism and directed towards machining zone. The combined effect of these mechanisms helps to increase the lubrication effect and the cooling effect.

Figure 10.2 depicts a MQCL system designed for milling operations. In this case, a dual-channeled system is used to combine the cold air and MQL aerosol before they reach the machining interface. Vegetable oil is commonly used for MQL

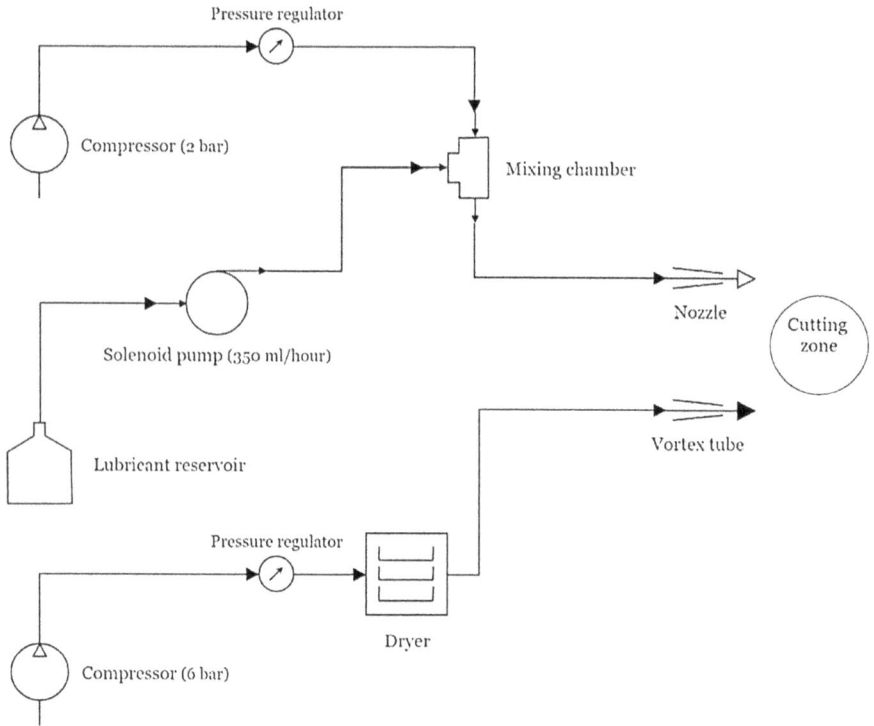

FIGURE 10.2   Schematic of the MQCL system (Mark Benjamin et al., 2018).

lubricating oil since it is biodegradable and has good lubricating characteristics. The MQL and the cold air outlet are located at a distance of 50 mm from the machining interface, which is the distance that has been determined to be ideal for MQL machining. The two outlets are directed towards the machining interface that moves along the tool and covers maximum of the cutting tool and the workpiece. The outlets or nozzles move along with the tool and provide the combination of cold air and aerosol (Mark Benjamin et al,. 2018) (Figure 10.3).

10.3   INFLUENCE OF MQCL ON DIFFERENT MACHINING PROCESSES

10.3.1   TURNING WITH MQCL

Cutting fluids are one of the many cooling techniques used to lower the cutting temperature at the machining interface during turning. The turning process generates a great deal of heat and results in severe cutting temperatures at the interface. But these cutting fluids affect the environment and create health hazards to the operator (Debnath et al,. 2014). In order to minimize these effects and promote sustainable machining, various cooling strategies have been introduced; among them, MQL shows promising results, which leads to a reduction in cutting temperature and heat

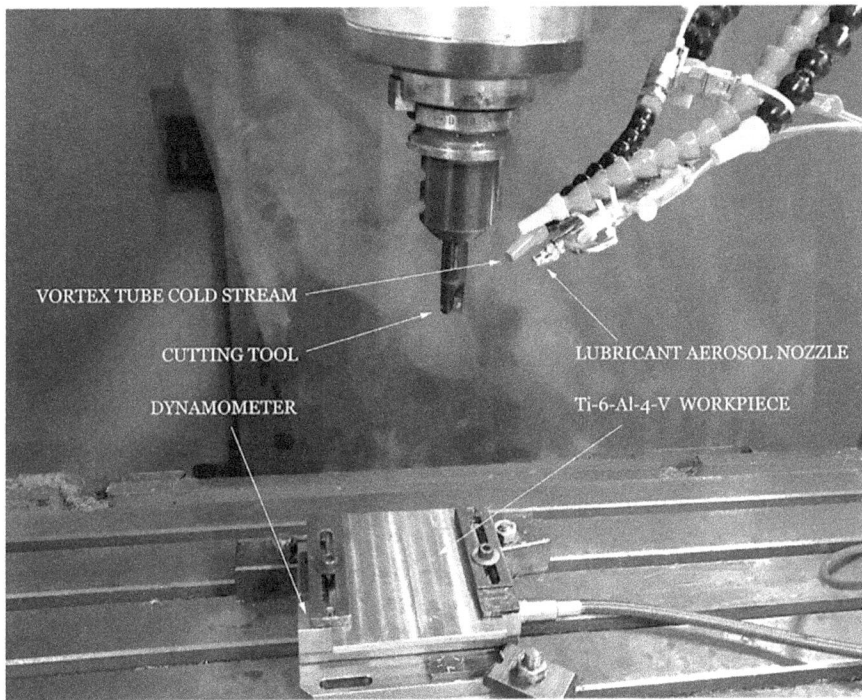

FIGURE 10.3   MQCL setup in vertical machining center (Mark Benjamin et al., 2018).

generation at the machining interface and moreover it also promotes sustainability as the lubricating oil (vegetable oil is biodegradable) (Race et al,. 2021). The cooling effect is one of the limitations of the MQL, so to overcome that, MQCL is introduced while turning titanium alloy and it was concluded that the MQCL technique can be viewed as a suitable alternative to conventional or synthetic cooling methods in terms of improving tool life and surface integrity characteristics. The combination of cooled air and lubrication while turning shows more effective results than dry machining. The author performed tests on cryogenic machining and found that the cryogenic cooling also helps in increasing machining performance even at a higher feed and speed. But in the case of cost, usage of cryogenic cooling can be compromised compared to MQCL (Raza et al,. 2014). Figure 10.4 represents the MQCL arrangement for the turning process where the MQCL nozzle is closer to and directed towards the machining interface, which helps in efficient lubrication and helps in removal of adequate amounts of heat from the machining interface. Liu et al. performed turning operations using flood cutting conditions and MQCL, as shown in Figure 10.5, and found that the vibrations in turning with MQCL were observed to be minimized compared to flood cutting conditions.

This is due to the influence of atomization of the lubricant oil. The resulting droplets were directed to the machining zone, where they provide appropriate lubrication and form oil films between the tool and the workpiece, lowering frictional force. Moreover, the cold air, which is in addition to the aerosol, enhances the heat

FIGURE 10.4 MQCL arrangement for turning operation (Liu et al., 2021).

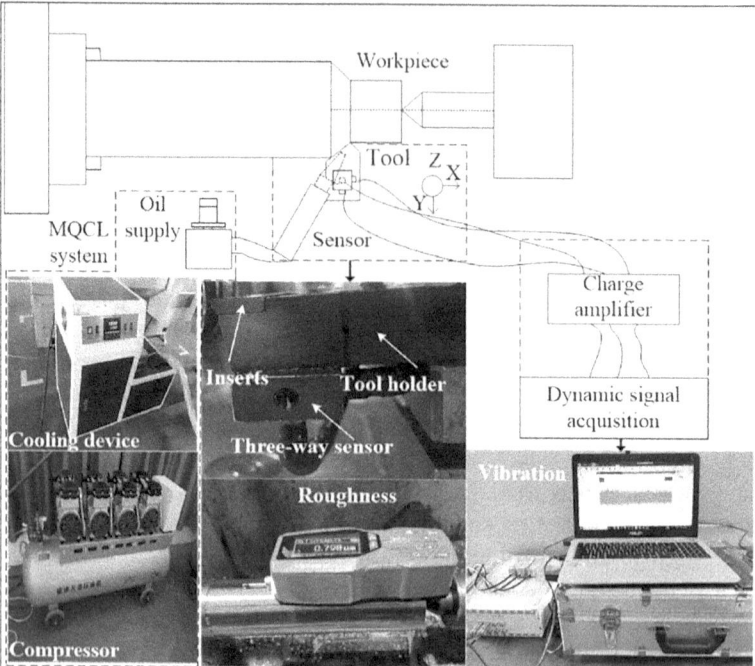

FIGURE 10.5 MQCL experimental setup (Liu et al., 2021).

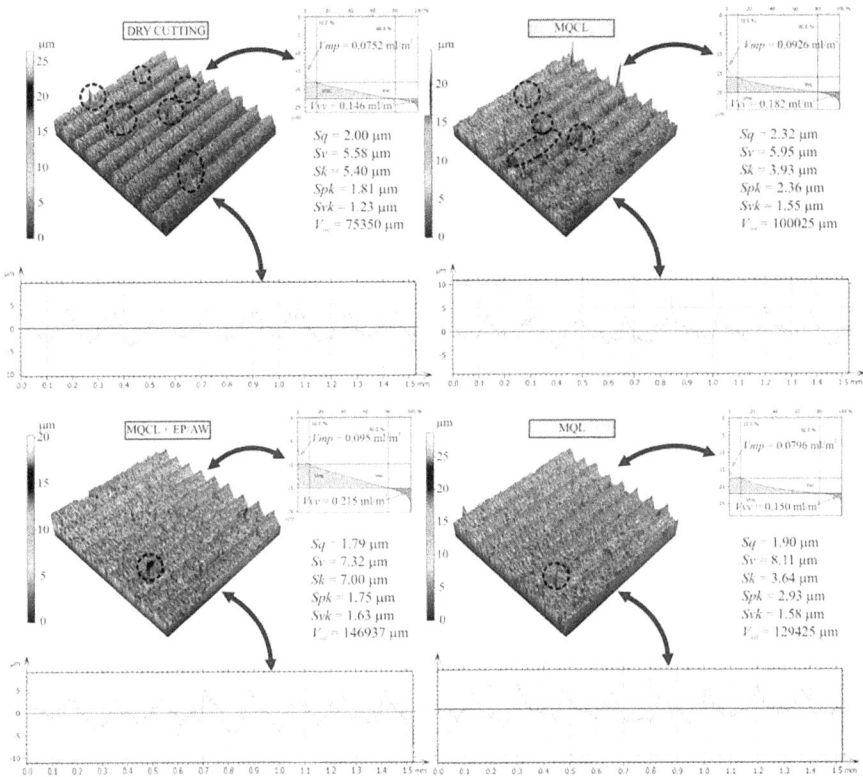

FIGURE 10.6 Surface topography of machined surfaces with dry, MQCL, MQCL+EP/AW, and MQL conditions (Maruda et al., 2020).

dissipation, resulting in a lower cutting force (Liu et al,. 2021). From Figure 10.6, we can see that the machined surface topography and the least values of S_q were found for MQCL+EP/AW, where the tribo-film formed is the reason for the oil retention and results in low surface roughness values. From Figure 10.7(a), the volumetric wear decreased significantly, which may be attributed to the generation of the tribo-film, where the wear of MQCL+EP/AW decreased by 16% when compared to the MQCL method and was minimized by 27% when compared to the dry machining conditions. From Figure 10.7(b), the temperature was low for MQCL+EP/AW, where a tribo-film thin layer was formed on the surface (Maruda et al,. 2020).

10.3.2 MILLING WITH MQCL

Krishnan et al. (2021) introduced multi-jets in MQL with cold air during the milling operation. Here, three jets are used where jet 1 and jet 3 are used for spraying aerosol and jet 2 is to spray aerosol and cold air. The orientation of jets and their distance from the machining interface are optimized to make sure the lubrication covers the entire machining operation. Here, the optimum position was found when jet 1 was kept at the angle of 20°, jet 2 kept at the angle of 90°, and jet 3 kept at the

(a)

(b)

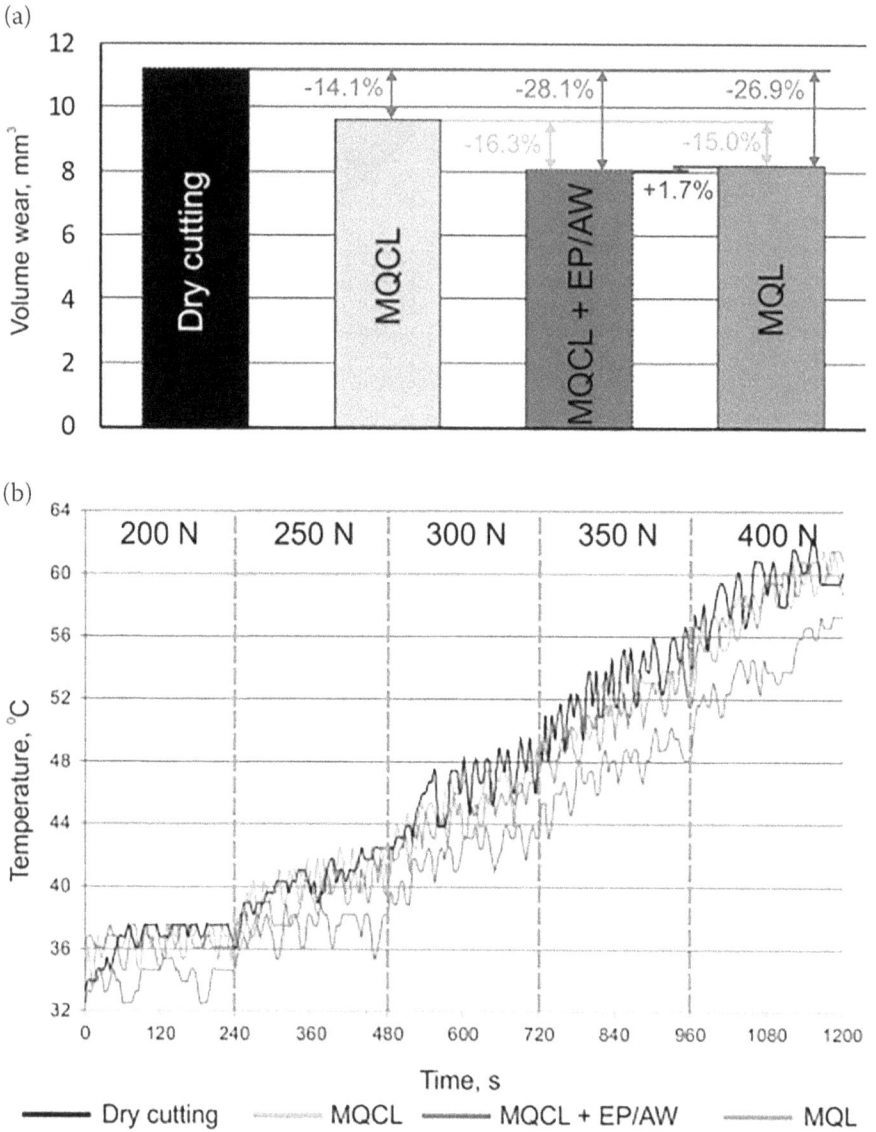

FIGURE 10.7   (a) Volumetric wear and (b) temperature at dry, MQCL, MQCL+EP/AW, MQL (Maruda et al., 2020).

angle of 160°. The current position results in 22% lower cutting force than other positions. When compared to dry, flood, and MQL conditions, the resultant force when using MJMQCL is 1.40, 1.60, and 1.75 times lower than MQL, flood, and dry conditions, respectively.

Heat is generated while milling is reduced as the cold air is supplied to the machining interface in addition to the MQL aerosol around the machining interface

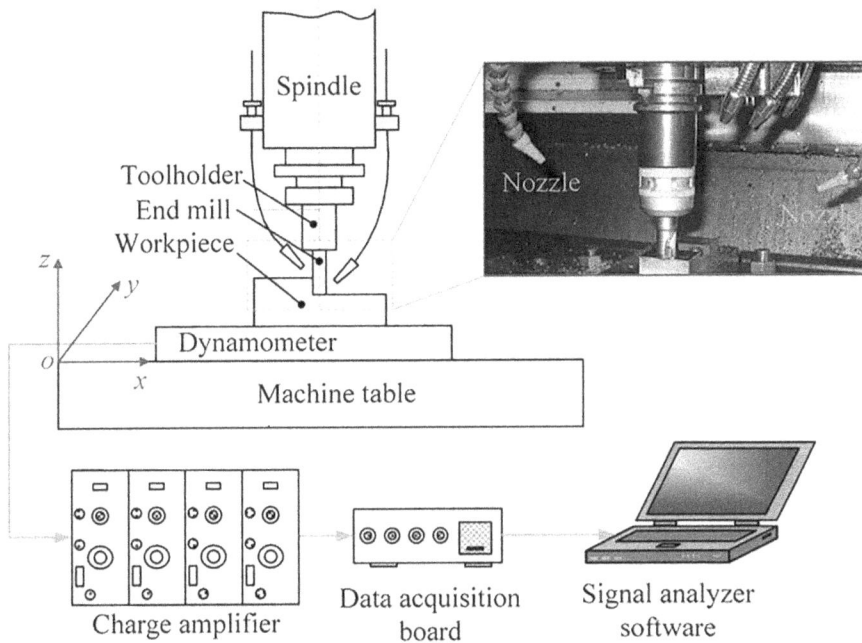

FIGURE 10.8 Experimental setup of MQCL milling Inconel 718 (Zhang et al., 2012).

in the case of MJMQCL. The temperature at the machining interface is minimal, ascribed to improvement in convective heat transfer coefficient and also an increase in the viscosity of the lubricating oil. The tool-chip machining interface temperature is reduced, ascribed to the film continuity that occurs in MJMQCL, which permits adequate time to transfer heat by droplet evaporation, which results in minimizing the temperature at the machining zone. Here, the position of the multi-jets also plays a major role in faster heat dissipation at the machining interface.

Zhang et al. (2012) observed the limitations of using MQL in milling Inconel 718 and opted to use MQCL to evaluate its benefits over MQL; here, the cold air is supplied to MQL at −35 °C and the oil used is synthetic vegetable oil, which is 100% biodegradable. Figure 10.8 shows the experimental setup for performing a milling operation of Inconel 718 with a MQCL setup. While the biodegradable vegetable oil was allowed to flow from one nozzle, the cold air was discharged from the other nozzle across the tool chip interfaces. Figure 10.9 shows the tool wear progression versus machining time at various cooling conditions. When MQCL is applied across the machining interface, the jet comes into contact with the workpiece and the tool at a very high velocity and at a high pressure. The authors found that the diameter of the oil determines the oil penetration across the machining interface. The smaller the size of the droplets, the easier it can be penetrated across the machining interface, which leads to the formation of oil film at these interfaces that assist in minimizing the friction. At this point, the cryogenic compressed air helps in bringing down the machining temperature across the interface. Attributed to the

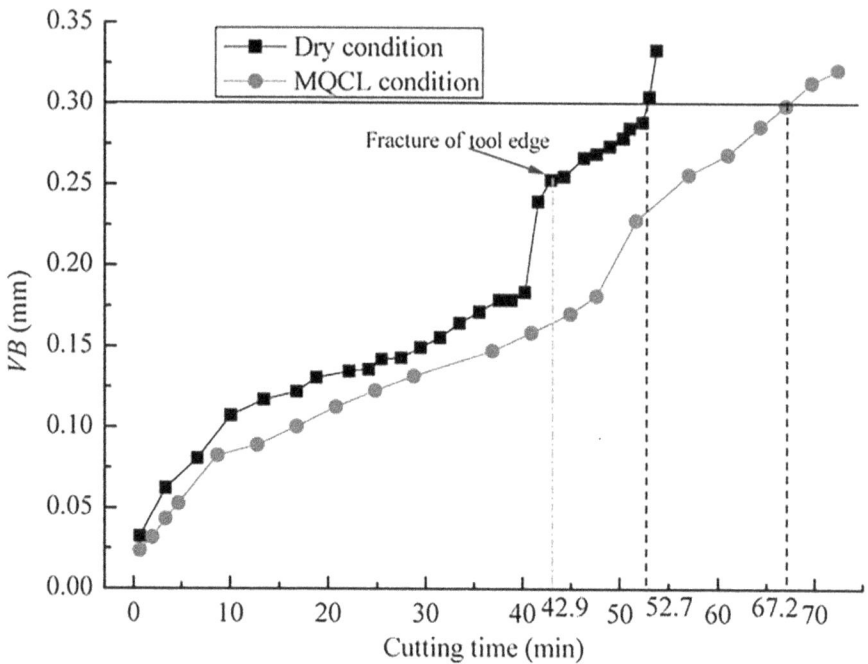

FIGURE 10.9   Growth of flank wear under dry and MQCL conditions (Zhang et al., 2012).

combined benefits of superior cooling and lubrication mechanism of the cryogenic compressed air and the vegetable oil, the tool life under the MQCL condition improved significantly in contrast to dry lubrication conditions.

The picture displays a comprehensive comparison of four different lubrication methods in terms of volumetric wear and temperature during machining processes. Lower bars indicate reduced wear, while higher bars signify greater wear on the cutting tools. Lower lines indicate better cooling efficiency and lower temperatures, while higher lines suggest increased heat generation during machining.

The picture illustrates the experimental setup for (MQCL) during the milling process of Inconel 718. In the image, a milling machine is prominently featured, equipped with the necessary tooling to perform precision milling on the Inconel 718 workpiece. The experimental setup may also feature advanced sensors and data acquisition systems to monitor various parameters during the milling process, such as cutting forces, temperature, and surface finish.

The picture showcases a comparison of the growth of flank wear on a cutting tool under two different lubrication conditions: dry machining and MQCL. The comparison between the two line graphs visually illustrates the benefits of employing MQCL over traditional dry machining. The reduced flank wear in the MQCL condition indicates improved cooling and lubrication at the cutting zone, resulting in enhanced tool life and potentially better surface finish.

Figure 10.10 illustrates the variation of machining forces under dry and MQCL conditions during end-milling of Inconel 718. At the initial condition, the edge of

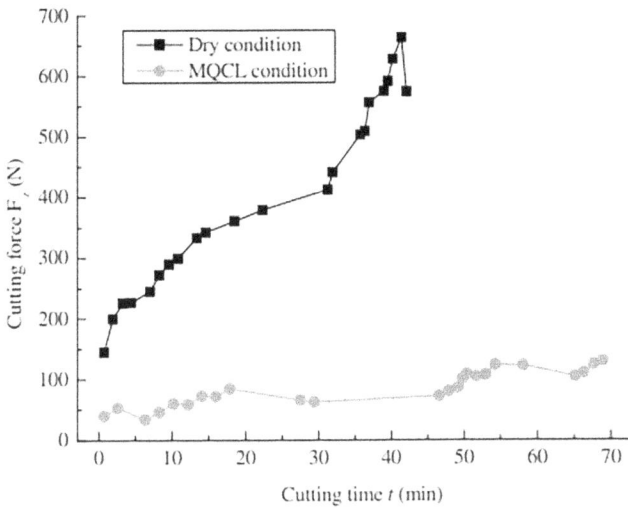

FIGURE 10.10 Variation of cutting force versus cutting time under dry and MQCL conditions (Zhang et al., 2012).

the tool is assumed to be very sharp because of the cutting force is low. However, on subsequent passes, the cutting edge loses its sharpness due to which high frictional forces may be observed. At the same time, machining with MQCL results in considerable reduction of wear due to tool sharpness being maintained for a longer duration of time, which reduces the friction and cutting forces (Zhang et al,. 2012).

10.3.3 GRINDING WITH MQCL

Grinding is commonly used as a finishing procedure for materials that are considered difficult to be machined. Grinding requires a significant amount of energy per unit volume for material to be removed. It then gets converted to heat energy, resulting in superior machining temperatures and workpiece damage such as cracks, accuracies, unfavorable residual stress, lower fatigue strength, as well as phase transformations that result in poor surface quality. To reduce these effects and improve the surface finish, Baumgart et al. (2017) was used MQCL in the grinding process. MQCL reduced the cutting forces compared to conventional cooling and MQL conditions, and it offers good grain sliding at the workpiece tool interface. Friction between the grain and workpiece is reduced due to the high fluid lubricating capacity in MQCL. Because grinding wheel wears less at the MQCL condition, the sharpness of the abrasive grains lasts longer, resulting in a high G-ratio. MQCL reduces roughness values more effectively than MQL and flood cooling due to efficient lubrication and low temperature or effective abrasive grain cooling effects near the tool-work contact interfaces.

In this case, the MQCL approach assists the chip in sliding easily across the tool face, thereby promoting a superior surface finish. After grinding using MQCL, no significant thermal burn was observed. Here, the MQCL technique leads to a very good surface finish compared to flood cutting conditions, due to significant oil penetration across the machining interface. MQCL gave lower surface roughness than MQL because of the effective cooling effect at the tool and workpiece interface (Li et al., 2010).

10.3.4 DRILLING WITH MQCL

The drilling process involves a large amount of heat accumulation and high friction at the contact zone, which reduces the effectiveness of the drilling operation. The operation involves the generation of superior machining temperatures and higher machining forces due to which the tool life and mainly surface profile and the roundness of the hole deteriorates. To minimize these effects and improve the life of the drill bit, MQCL was used in the hard drilling of Hardox 500 steel (Duc et al., 2020). The cold air of MQCL is produced from the Frigid-X Sub-Zero Vortex Tool Cooling Mist System. The authors used rice bran oil as a lubricating oil. In comparison with MQL, MQCL demonstrated a significant improvement in surface roughness values. Lubricating oil, which comes between the tool and the workpiece, reduces friction by easy removal of chips from the machining interface. Additionally, cold air from the outlet helps in minimizing the machining temperatures. Cold air enhances heat dissipation, which helps in reduction of heat accumulation at the machining interface.

MQCL condition has shown a noticeable reduction in tool wear, especially flank wear. The tool life is increased to 90 min, while the drilling under dry conditions shows the drill bit tool life is 20 min at a machining speed of 20 m/min and the feed is 0.04 mm/rev. Drilling thrust force is low for MQCL cutting conditions compared to dry and MQL as the lubrication reduces the friction, which reduces the force and the increased tool life results in a prolonged sharp edge while drilling. The cutting edge, which is in direct contact with the work specimen, is not damaged during dry cutting and MQL condition. As the tool wear is reduced, the distortions are also observed to be minimal, due to which thrust forces are observed to be significantly reduced for the MQCL condition. The chip flow is increased as the lubrication enhances the slippery action between the chip and the tool. Moreover, the adhesion of the tool was also observed to be less for the MQCL condition, due to enhanced heat dissipation and the lubrication mechanism.

10.4 INFLUENCE OF ADDITIVES ON THE PERFORMANCE OF MQCL

The nanofluids suspended in MQL have proven to offer an alternate option for hard-to-cut materials while being environmentally benign. From this concept, the utilization of nanoparticles in MQCL can help in increasing machining performance for hard-to-cut materials and also promote sustainability. MoS_2, one of the nanoparticles used in MQCL, shows promising results, where the results showed that the surface roughness (R_a) values were observed to be significantly lower than the dry and MQL lubrication conditions. Surface roughness is improved after utilizing MoS_2 concentrations of 0.2 and 0.5 wt% in comparison to MQCL with the fluid, but the roughness values increased for the concentration of 0.8 wt% due to MoS_2 nanoparticles are ellipsoidal, which have a low friction coefficient. The white layer is reduced significantly due to the efficient cooling and the better lubrication for the MQCL environment when nanoparticles are added (Dong et al., 2019).

Tribo-film formation is observed while using MoS_2, which is called *micro-bubbles*. The microbubbles can be observed on a machined surface and are responsible for the better lubrication mechanism. Tuan et al. (2022) investigated the impacts of hybrid nanofluids used in MQCL during hard milling of Hardox 500 steel and assessed the surface integrity of the machined specimen. Al_2O_3 improves the thermal conductivity of the base fluid, which aids in the formation of the ball roller effect. However, MoS_2 has the effect of increasing tribo-film and helps to decrease the friction at the machining interface by forming a tribo-film. Hybrid lubrication in MQCL produces a good surface finish (Duc et al., 2021). Graphene plates are used in MQCL to enhance the cooling and lubrication at the machining interface. Graphene platelets have a very high thermal conductivity, which increases the cooling capacity of the lubricant oil.

These nanoparticles, when delivered at the machining interface, settle across the tool-chip interface and assist chips in easily moving out of the machining interface, due to which the friction across the tool-chip interface is significantly reduced. Furthermore, nanoparticles mixed with lubricating oil increase the heat dissipation capacity at the machining interface, lowering the machining interface temperature, and tool-chip contact friction. Due to these advantages, MQCL with GnP decreases the cutting forces on the tool by reducing the friction as the GnP helps in increasing the sliding action across the tool-chip interfaces. Figure 10.11 shows the tool wear images of different cutting conditions of dry, MQCL, and MQCL oil-GnP. From this figure we can see the wear is reduced for MQCL and MQCL oil-GnP. This is ascribed to the significantly minimized friction across the tool-chip interfaces.

GnP, delivered across the tool-chip interfaces, reduces friction, resulting in less adhesion and abrasion wear. Here, the temperature is also less when compared to the dry and MQCL conditions, due to high thermal conductivity of GnP, which helps in effective dissipation of machining heat from the contact interface. Reduction in temperature leads to an effective decrease in adhesion wear. Due to the combined effect of MQCL and GnP, the tool life is enhanced significantly (Gutnichenko et al., 2018).

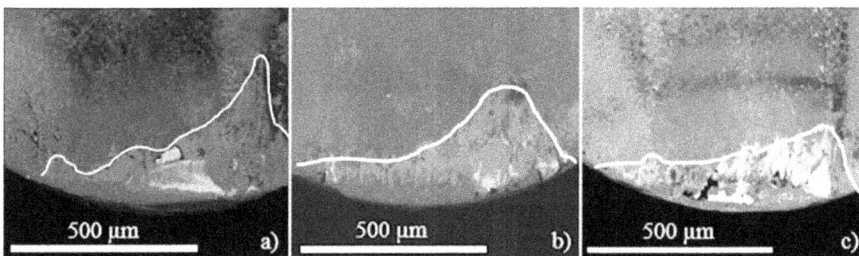

FIGURE 10.11 Tool wear images of a) dry, b) MQCL, and c) MQCL-GnP (Gutnichenko et al., 2018).

10.5 CONCLUSION

In this chapter, details about a new cooling or lubrication strategy is discussed. The benefits of the cold air and the lubricant used as a combined lubrication medium has been discussed on various machining processes.

1. When compared to dry machining, MQCL shows effective results for finish turning of low-carbon steels due to the easy removal of chips from the machining interface and the surface integrity of the machined specimen is improved considerably when compared to compressed air cooling and dry machining lubrication conditions.
2. The cooling air and minimum quantity lubrication (CAMQL) resulted in a significant decrease in tool wear and improved tool life in the milling of AISI D2, while the chips obtained were observed to be curly in nature, which is an indication of their reduced tool-chip contact length.
3. A thin layer of tribo-film is generated across the tool surface while using MQCL, which reduces the coefficient of friction values across the tool and chip interfaces. The tribo-film also reduces adhesion-diffusion and chemical wear. Here, the droplets, which have smaller diameters, penetrate into the machining interface and can easily evaporate.
4. Nanoadditives used in MQCL enhance the machining performance as they improve the cooling and lubricating effects that can be utilized in machining difficult-to-cut materials. The use of different types of lubricating oils by suspending nanoparticles shows promising results in improving the machining performance of hard-to-machine work specimens. Nanoparticles, which have a very high thermal conductivity, increase the heat dissipation at the machining interface, thereby bringing dowing the machining temperatures significantly.
5. The intervention of cold air or sub-zero air in the minimum quantity lubrication system results in lowering the frictional coefficient values at the tool-rake interface. The combined effect of the hybrid system results in minimizing the temperatures while also assisting in easy separation of the chips, which results in a better surface finish and enhanced tool life features. The effect of lubricating action of the lubricant and also chip separation caused by the bimetallic spring effect formed by cold air and enhanced heat transfer coefficient resulted in promoting a better machining performance than at MQL conditions.

REFERENCES

Achparaki, M., Thessalonikeos, E., Tsoukali, H., Mastrogianni, O., Zaggelidou, E., Chatzinikolaou, F., Vasilliades, N., Raikos, N., Isabirye, M., Raju, D. V., Kitutu, M., Yemeline, V., Deckers, J., & J. Poesen Additional. (2012). We Are IntechOpen, the World' s Leading Publisher of Open Access Books Built by Scientists, for Scientists TOP 1%. *Intech*, 13.

Baumgart, C., Radziwill, J. J., Kuster, F., & Wegener, K. (2017). A study of the interaction between coolant jet nozzle flow and the airflow around a grinding wheel in cylindrical grinding. *Procedia CIRP*, 58, 517–522.

Debnath, S., Reddy, M. M., & Yi, Q. S. (2014). Environmental friendly cutting fluids and cooling techniques in machining: A review. *Journal of Cleaner Production*, 83, 33–47.

Dong, P. Q., Duc, T. M., & Long, T. T. (2019). Performance evaluation of Mqcl hard milling of Skd 11 Tool steel using Mos2 Nanofluid. *Metals (Basel).*, 9(6).

Duc, T. M., Long, T. T., & Tuan, N. M. (2021). Novel uses of Al2O3/MoS2 hybrid nanofluid in MQCL hard milling of Hardox 500 Steel. *Lubricants*, 9(4).

Duc, T. M., Long, T. T., & Van Thanh, D. (2020). Evaluation of minimum quantity lubrication and minimum quantity cooling lubrication performance in hard drilling of hardox 500 steel using Al2O3 nanofluid. *Advances in Mechanical Engineering*, 12(2), 1–12.

Gutnichenko, O., Bushlya, V., Bihagen, S., & Ståhl, J. E. (2018). Influence of graphite nanoadditives to vegetable-based oil on machining performance when MQCL assisted hard turning. *Procedia CIRP*, 77, 437–440.

Krishnan, G. P., P, S., Samuel Raj, D., Hussain, S., Ravi Shankar, V., & Raj, N. (2021). Optimization of jet position and investigation of the effects of multijet MQCL during end milling of Ti-6Al-4V. *Journal of Manufacturing Processes*, 64, 392–408.

Li, C., Du, C., Liu, G., & Zhou, Y. (2010). Performance evaluation of minimum quantity cooling lubrication using CBN grinding wheel. *Advanced Materials Research*, 97–101, 1827–1831.

Liu, N., Liu, B., Jiang, H., Wu, S., Yang, C., & Chen, Y. (2021). Study on vibration and surface roughness in MQCL turning of stainless steel. *Journal of Manufacturing Processes*, 65, 343–353.

Mark Benjamin, D., Sabarish, V. N., Hariharan, M. V., & Samuel Raj, D. (2018). On the benefits of sub-zero air supplemented minimum quantity lubrication systems: An experimental and mechanistic investigation on end milling of Ti-6-Al-4-V Alloy. *Tribology International*, 119, 464–473.

Maruda, W. R., Legutko, S., & Krolczyk, G., (2014). Effect of minimum quantity cooling lubrication (MQCL) on chip morphology and surface roughness in turning low carbon steels. *Applied Mechanics and Materials*, 657, 38–42.

Maruda, R. W., Krolczyk, G. M., Feldshtein, E., Pusavec, F., Szydlowski, M., Legutko, S., & Sobczak-Kupiec, A. (2016). A study on droplets sizes, their distribution and heat exchange for minimum quantity cooling lubrication (MQCL). *International Journal of Machine Tools and Manufacture*, 100, 81–92.

Maruda, R. W., Krolczyk, G. M., Feldshtein, E., Nieslony, P., Tyliszczak, B., & Pusavec, F. (2017). Tool wear characterizations in finish turning of AISI 1045 carbon steel for MQCL conditions. *Wear*, 372–373, 54–67.

Maruda, R. W., Krolczyk, G. M., Wojciechowski, S., Powalka, B., Klos, S., Szczotkarz, N., Matuszak, M., & Khanna, N. (2020). Evaluation of Turning with different cooling-lubricating techniques in terms of surface integrity and tribologic properties. *Tribology International*, 148, 106334.

Okokpujie, I. P., Tartibu, L. K., Sinebe, J. E., Adeoye, A. O. M., & Akinlabi, E. T. (2022). Comparative Study of rheological effects of vegetable oil-lubricant, TiO2, MWCNTs nano-lubricants, and machining parameters influence on cutting force for sustainable metal cutting Process. *Lubricants*, 10(4).

Pal, A., Chatha, S. S., & Sidhu, H. S. (2021). Performance evaluation of the minimum quantity lubrication with Al_2O_3- mixed vegetable-oil-based cutting fluid in drilling of AISI 321 stainless steel. *Journal of Manufacturing Processes*, 66, 238–249.

Race, A., Zwierzak, I., Secker, J., Walsh, J., Carrell, J., Slatter, T., & Maurotto, A. (2021) Environmentally sustainable cooling strategies in milling of SA516: Effects on surface integrity of dry, flood and mql machining. *Journal of Cleaner Production*, 288, 125580.

Raza, S. W., Pervaiz, S., & Deiab, I. (2014). Tool wear patterns when turning of titanium alloy using sustainable lubrication strategies. *International Journal of Precision Engineering and Manufacturing*, 15(9), 1979–1985.

Su, Y., He, N., Li, L., Iqbal, A., Xiao, M. H., Xu, S., & Qiu, B. G. (2007). Refrigerated cooling air cutting of difficult-to-cut materials. *International Journal of Machine Tools and Manufacture*, 47(6), 927–933.

Subhedar, D. G., Patel, Y. S., Ramani, B. M., & Patange, G. S. (2021). An experimental investigation on the effect of Al2O3/ cutting oil based nano coolant for minimum quantity lubrication drilling of SS 304. *Cleaner Engineering and Technology*, 3, 100104.

Tuan, N. M., Duc, T. M., Long, T. T., Hoang, V. L., & Ngoc, T. B. (2022). Investigation of machining performance of MQL and MQCL hard turning using nano cutting fluids. *Fluids*, 7(5).

Zhang, S., Li, J. F., & Wang, Y. W. (2012). Tool life and cutting forces in end milling Inconel 718 under dry and minimum quantity cooling lubrication cutting conditions. *Journal of Cleaner Production*, 32, 81–87

Index

For Product Safety Concerns and Information please contact our EU
representative GPSR@taylorandfrancis.com
Taylor & Francis Verlag GmbH, Kaufingerstraße 24, 80331 München, Germany

www.ingramcontent.com/pod-product-compliance
Lightning Source LLC
Chambersburg PA
CBHW070717220326
41598CB00024BA/3194